# COMPUTATIONAL FLUID DYNAMICS AND COMSOL MULTIPHYSICS

*A Step-by-Step Approach for Chemical Engineers*

# COMPUTATIONAL FLUID DYNAMICS AND COMSOL MULTIPHYSICS

*A Step-by-Step Approach for Chemical Engineers*

Ashish S. Chaurasia, PhD

First edition published [2022]

**Apple Academic Press Inc.**
1265 Goldenrod Circle, NE,
Palm Bay, FL 32905 USA

4164 Lakeshore Road, Burlington,
ON, L7L 1A4 Canada

**CRC Press**
6000 Broken Sound Parkway NW,
Suite 300, Boca Raton, FL 33487-2742 USA

2 Park Square, Milton Park,
Abingdon, Oxon, OX14 4RN UK

**Library and Archives Canada Cataloguing in Publication**

Title: Computational fluid dynamics and COMSOL Multiphysics : a step-by-step approach for chemical engineers / Ashish S. Chaurasia, PhD.
Names: Chaurasia, Ashish S., author.
Description: First edition. | Includes bibliographical references and index.
Identifiers: Canadiana (print) 20210290668 | Canadiana (ebook) 20210291036 | ISBN 9781774630082 (hardcover) | ISBN 9781774639320 (softcover) | ISBN 9781003180500 (PDF)
Subjects: LCSH: Computational fluid dynamics. | LCSH: Computational fluid dynamics—Computer simulation—Handbooks, manuals, etc. | LCSH: COMSOL Multiphysics—Handbooks, manuals, etc. |
LCGFT: Handbooks and manuals.
Classification: LCC TA357.5.D37 C43 2021 | DDC 620.1/064028553—dc23

**Library of Congress Cataloging-in-Publication Data**

Names: Chaurasia, Ashish S., 1976- author.
Title: Computational fluid dynamics and COMSOL Multiphysics : a step-by-step approach for chemical engineers / Ashish S. Chaurasia, PhD.
Description: First edition. | Palm Bay : Apple Academic Press, 2022. | Includes bibliographical references and index. | Summary: "This textbook, Computational Fluid Dynamics and COMSOL Multiphysics: A Step-by-Step Approach for Chemical Engineers, covers computational fluid dynamics simulation using COMSOL Multiphysics® Modeling Software in chemical engineering applications. In the volume, the COMSOL Multiphysics package is introduced and applied to solve typical problems in chemical reactors, transport processes, fluid flow, and heat and mass transfer. Inspired by the difficulties of introducing the use of COMSOL Multiphysics software during classroom time, the book incorporates the author's experience of working with undergraduate, graduate, and postgraduate students, which helps to make the book user friendly and that, at the same time, addresses typical examples within the subjects covered in the chemical engineering curriculum. Real-world problems require the use of simulation and optimization tools, and this volume shows how COMSOL Multiphysics software can be used for that purpose. The book includes over 560 snapshots to provide a step-by-step approach to using the program for computational fluid dynamics simulations. Key features: Includes step-by-step screenshots for all the examples Shows the graphical user interface of COMSOL, which does not require any programming effort Provides chapter-end problems for extensive practice along with solutions Includes actual examples of chemical reactors, transport processes, fluid flow, and heat and mass transfer This book is intended for students who want or need help to solve chemical engineering assignments using computer software. It can also be used for computational courses in chemical engineering. It will also be a valuable resource for professors, research scientists, and practicing engineers"-- Provided by publisher.
Identifiers: LCCN 2021039183 (print) | LCCN 2021039184 (ebook) | ISBN 9781774630082 (hardback) | ISBN 9781774639320 (paperback) | ISBN 9781003180500 (ebook)
Subjects: LCSH: Computational fluid dynamics. | Fluid dynamics. | COMSOL Multiphysics.
Classification: LCC TA357.5.D37 C43 2022 (print) | LCC TA357.5.D37 (ebook) | DDC 620.1/064--dc23
LC record available at https://lccn.loc.gov/2021039183
LC ebook record available at https://lccn.loc.gov/2021039184

ISBN: 978-1-77463-008-2 (hbk)
ISBN: 978-1-77463-932-0 (pbk)
ISBN: 978-1-00318-050-0 (ebk)

# About the Author

**Ashish S. Chaurasia, PhD**

Ashish S. Chaurasia, PhD, teaches in the Chemical Engineering Department at Visvesvaraya National Institute of Technology, where he is an Associate Professor. He has more than 19 years of teaching and research experience. He received the PhD degree in chemical engineering from the Birla Institute of Technology and Science in 2004. He was a postdoctoral fellow at the National Chemical Laboratory, India, from 2005 to 2006 and a postdoctoral research associate at Imperial College London, United Kingdom, from 2006 to 2008. His general research interests focus on pyrolysis of biomass, biomass gasification, computational fluid dynamics, and modeling and simulation. He has authored two books and more than 100 publications in reputed journal and conferences.

# Contents

# Abbreviations

CFD   computational fluid dynamics
CSTR   continuous stirred tank reactor
PDE   partial differential equation

# Preface

This textbook covers computational fluid dynamics simulation using Comsol Multiphysics software. It is assumed that the institute/university using this book has access to this software. The fifteenday trial version of the software can be downloaded at https://www.comsol.co.in/product-download.

This book is suitable for professors, instructors, and undergraduate and postgraduate students of chemical engineering. It will also be helpful to research scientists and practicing engineers. This book can also be used for longer duration computational courses in chemical engineering such as Chemical Engineering Mathematics, CFD for Chemical Engineers, Introduction to Softwares in Chemical Engineering, Computational Design Laboratory, Computer Applications in Chemical Engineering Laboratory, Numerical Methods, etc.

This book can also be used for short duration in chapters for different courses such as Chemical Reaction Engineering, Heat Transfer, Mass Transfer, Transport Phenomenon, etc. Once the particular chapter is discussed in the course, the students can perform the laboratory sessions in Comsol for detailed understanding of the concept.

The book provides a step-by-step approach to solve the problems with snapshots that can be understood very easily. The book is inspired by the difficulties while introducing the use of Comsol within class hours. It addresses the need for a textbook that is user friendly and that, at the same time, addresses typical examples within the subjects covered in the chemical engineering curriculum. The aim of this volume, therefore, is to provide the use of Comsol Multiphysics software in chemical engineering applications so that students become familiar with the capabilities.

This book solves typical problems in chemical reactors, transport processes, fluid flow, heat and mass transfer using most popular commercial Comsol Multiphysics software. It also discusses the real-world problems that require the use of simulation and optimization tools. The specific highlights of this book include:

1. Use of a graphical user interface of COMSOL which does not require any programming effort.
2. Includes actual examples on chemical reactors, transport processes, fluid flow, heat and mass transfer.

3.  The textbook is user friendly as it includes step by step screen shots for all the examples.
4.  Provides chapter-end problems for extensive practice.
5.  Provides solutions of all the problems in CD.

I would like to acknowledge the class interaction with undergraduate, graduate, and postgraduate students, which has greatly contributed toward the shaping of this book. I have also been benefited by the lively discussions with Mr. Rohit More, who is professional at Comsol Multiphysics Pvt Ltd, Bengaluru, India.

I acknowledge the support of the Apple Academic Press team. I am also grateful to the anonymous reviewer whose valuable comments and suggestions for improvement have gone a long way in the formation of the final version of this book.

I would also like to acknowledge the contribution of the Visvesvaraya National Institute of Technology, Nagpur, India, for providing necessary facilities to carry out this work.

The author appreciates any comment and suggestion from readers through email.

—**Ashish S. Chaurasia, PhD**
Department of Chemical Engineering
Visvesvaraya National Institute of Technology, Nagpur
aschaurasia@che.vnit.ac.in

# CHAPTER 1

# Introduction

The past three decades have witnessed the phenomenal growth in the area of computational fluid dynamics (CFD) due to the developments in the field of computers. CFD has now become the integral part the way engineers design and analyze the processes. The engineers can make use of CFD tools to simulate the reaction kinetics, fluid flow, and heat transfer processes in the design and predict the process performance before production. The advantages of using CFD tools include very high speed of computing, much cheaper than setting up big experiments or building prototypes, allows numerical experimentation, and so on.

The aim of this book is to provide the use of COMSOL Multiphysics software in chemical engineering applications so that students become familiar with the capabilities. This book solves typical problems in chemical reactors, transport processes, fluid flow, and heat and mass transfer using most popular commercial COMSOL Multiphysics software. It also discusses the real-world problems that require the use of simulation and optimization tools. The book is organized into seven chapters.

Chapter 2 deals with methods to model chemical reactors using differential and algebraic equations:

$$u\frac{dC_A}{dz} = -2kC_{A,s}^2, u\frac{dC_B}{dz} = +kC_{A,s}^2, u\frac{dC_C}{dz} = 0 \tag{1.1}$$

$$C_A(0) = 2\,\mathrm{kmol\,/\,m^3}, C_B(0) = 0, C_C(0) = 2\,\mathrm{kmol\,/\,m^3}. \tag{1.2}$$

In the differential equation (1.1), the dependent variables $C_A$, $C_B$ and $C_C$ are functions of independent variable $z$. The concentrations of different species in the reactor can be determined using initial conditions given in Equation (1.2)

$$148.4 = \frac{(x/2)(x/2)}{[(1-x)/2][(1-x)/2]}. \tag{1.3}$$

Equation (1.3) is nonlinear and can be solved by taking an appropriate initial guess value of $x$

$$120c1 - 20c2 = 400 \tag{1.4}$$

$$c1 - c2 = 0 \tag{1.5}$$

$$40c1 + 60c2 - 120c3 = -200. \tag{1.6}$$

Equations (1.4)–(1.6) represent the linear algebraic equations for three reactors linked by the pipes and can be solved for concentrations $c1$, $c2$, and $c3$.

Chapter 3 deals with problems in transport processes in one dimension considering steady and transient state. Consider a hot infinite plate of finite thickness having insulation condition on the left boundary and dimensionless temperature $\theta = 0$ at the other boundary with initial condition $\theta(X, 0) = 1$. The governing differential equation is

$$\frac{\partial \theta}{\partial \tau} = \alpha \frac{\partial^2 \theta}{\partial X^2} \tag{1.7}$$

where, $a = 2$, $\theta = \dfrac{T - T_\infty}{T_i - T_\infty}$, $X = \dfrac{x}{L}$, $\tau = \dfrac{\alpha t}{L^2}$

Equation (1.7) is a partial differential equation with a dependent variable $\theta$, which varies with two independent variables $X$, dimensionless distance, and $\tau$, dimensionless time. This partial differential equation can be solved easily using COMSOL Multiphysics.

Chapters 4 and 5 demonstrate the application of COMSOL Multiphysics to solve fluid flow and heat and mass transfer problems in two and three dimensions. Consider the flow of Newtonian fluid into a pipe described mathematically by the Navier–Stokes equations (1.8) and (1.9) and the continuity equation (1.10) for 2D, laminar, incompressible flow with constant viscosity

$$\rho \left( \frac{\partial u}{\partial t} + u \frac{\partial u}{\partial r} + w \frac{\partial u}{\partial z} \right) = -\frac{\partial p}{\partial r} + \mu \left( \frac{\partial^2 u}{\partial r^2} + \frac{\partial^2 u}{\partial z^2} \right) \tag{1.8}$$

$$\rho \left( \frac{\partial w}{\partial t} + u \frac{\partial w}{\partial r} + w \frac{\partial w}{\partial z} \right) = -\frac{\partial p}{\partial z} + \mu \left( \frac{\partial^2 w}{\partial r^2} + \frac{\partial^2 w}{\partial z^2} \right) \tag{1.9}$$

$$\frac{\partial u}{\partial r} + \frac{\partial w}{\partial z} = 0. \tag{1.10}$$

Equations (1.8)–(1.10) are partial differential equations with dependent variables $u$ and $w$, which vary with three independent variables $t$, $r$, and $z$. These partial differential equations require four boundary conditions and one initial condition to solve. Such type of complex real-life problems require power of software and can be solved easily using COMSOL Multiphysics.

Chapter 6 discusses the application of *COMSOL* to solve optimization problems using the *Optimization* module. The examples discussed include optimal cooling of a tubular reactor, optimization of a catalytic microreactor, and use of optimization in determination of Arrhenius parameters. Chapter 7 discusses the case studies on modeling, simulation, and optimization of the downdraft gasifier.

# CHAPTER 2

# Chemical Reactors

## 2.1 INTRODUCTION

The implementation of any industrial chemical process involves the use of several pieces of equipment interconnected into a chemical processing unit or plant. Chemical reactors are key pieces in the overall plant layout. Chemical reactors intended for use in different processes differ in design, size, and geometry. For all the differences, however, reactors may be classified on the basis of common features so that data can be presented in a systematic way. The reactors can be described mathematically, and the choice of the design procedures can be made. Chemical reactors differ in the number of features that affect their design and construction. In this chapter, several chemical reactor problems are solved using COMSOL Multiphysics.

## 2.2 SIMULATION OF A SIMPLE DIFFERENTIAL EQUATION

### 2.2.1 PROBLEM STATEMENT

Solve the following differential equation:

$$\frac{dy}{dt} = -10y. \tag{2.1}$$

$$\text{At } t = 0, \, y = 1, \tag{2.2}$$

compute the value of $y$ at $t = 1$ using a step size of $0.01$.

### 2.2.2 SIMULATION APPROACH

*Step 1*

- Select *0D Space Dimension* from the list of options by opening the COMSOL Multiphysics. Hit the *next* arrow at the upper right corner.

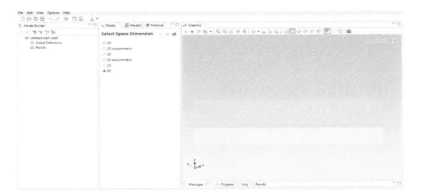

- Select and expand the *Mathematics* folder from the list of options in *Model Wizard*. Further choose *Global ODEs and DAEs* and hit *next tab*.

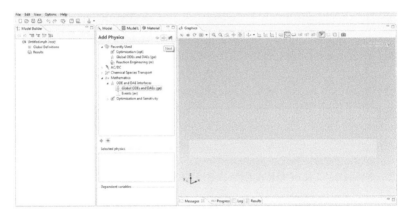

- Then, select *Time Dependent* from the list of *Study Type* options and click on the *Finish flag* at the upper right corner of the application.

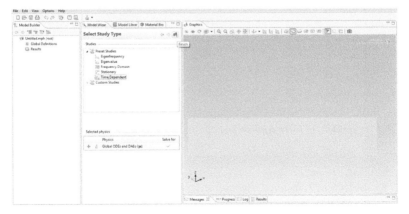

*Step 2*

- Now, in the *Model Builder*, select *Global Equations* option. This will open a tab to enter the coefficients of the characteristics of a model equation.

- In the *Global Equations* panel, you will see an equation of the form

$$f\left(u,u_t,u_{tt},t\right) = 0, u\left(t_0\right) = u_0, u_t\left(t_0\right) = u_{t0} \tag{2.3}$$

To convert Equation (2.1) to the desired form of Equation (2.3), enter the values as given in the following table.

| Name | $f(u, u_t, u_{tt}, t)$ | Initial value $(u\_0)$ | Initial value $(u\_t0)$ | Description |
|------|------------------------|------------------------|-------------------------|-------------|
| $y$  | $yt + 10y$             | 1                      | 0                       |             |

This indicates that the variable $y$ has an initial value of 1 at $t = 0$. In COMSOL, $\frac{dy}{dt} + 10y$ can be written as $yt + 10y$.

This adjustment will convert Equation (2.1) to the desired form of Equation (2.3).

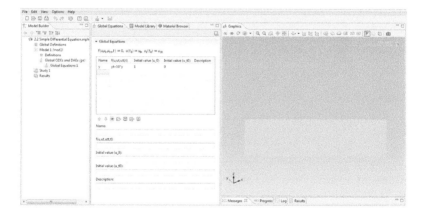

*Step 3*

- Now, expand the *Study 1* tab in *Model Builder*. Select *Step 1: Time Dependent*. Another window will open namely *Time Dependent*. Expand the *Study Settings* tab and set the range *Times: range (0,0.01,1)*.

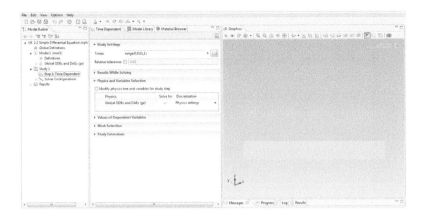

- Click on the *Compute (=)* button. A plot of *y* versus *t* appears as shown in Figure 2.1.
- Save the simulation.

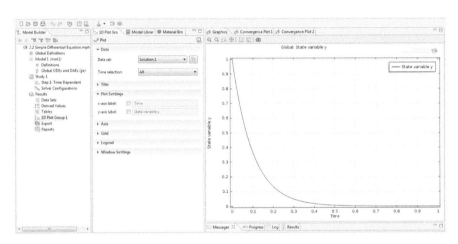

**FIGURE 2.1**    Plot of *y* as a function of time (Solution to Equation 2.1).


## 2.3    SIMULATION OF PLUG FLOW REACTOR

### *2.3.1    PROBLEM STATEMENT*

The following reaction takes place in a plug flow reactor (Fogler, 2006):

$$A \xrightarrow{k_1} B \xrightarrow{k_2} C.$$

Equation for components *A*, *B,* and *C* are

$$u\frac{dC_A}{dz} = -k_1 C_A \tag{2.4}$$

$$u\frac{dC_B}{dz} = k_1 C_A - k_2 C_B \tag{2.5}$$

$$u\frac{dC_C}{dz} = k_2 C_B \tag{2.6}$$

where
$u$ = velocity;
$z$ = distance from the inlet;
$C_i$ = molar concentration of the *i*th species.

At the inlet:

$$C_A(0) = 1\,\text{kmol}/\text{m}^3, C_B(0) = 0, C_C(0) = 0\,\text{kmol}/\text{m}^3 \tag{2.7}$$

and we take $u = 1\,\text{m}/\text{s}$, $k_1 = 1\text{s}^{-1}$, $k_2 = 0.1\text{s}^{-1}$, and reactor length as $z = 5$ m.

### 2.3.2  SIMULATION APPROACH (METHOD 1)

*Step 1*

- Select *0D Space Dimension* from the list of options by opening the COMSOL Multiphysics. Hit the *next* arrow at the upper right corner.

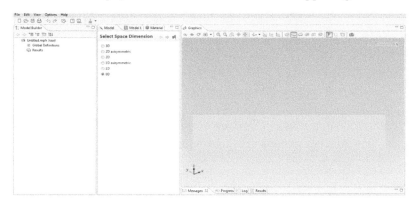

- Select and expand the *Mathematics* folder from the list of options in *Model Wizard*. Further choose *Global ODEs and DAEs* and hit *next* tab.

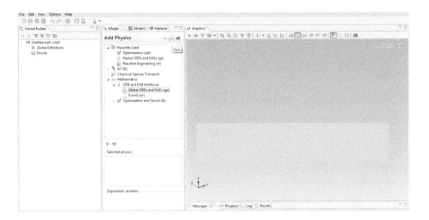

- Then, select *Time Dependent* from the list of *Study Type* options, and click on the *Finish flag* at the upper right corner of the application.

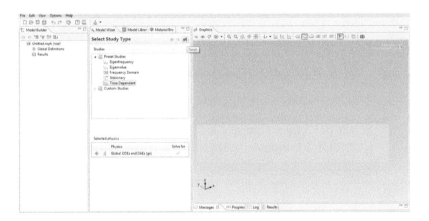

*Step 2*

- Now, in the *Model Builder*, select *Global Equations* option. This will open a tab to enter the coefficients of the characteristics of a model equation.
- In the *Global Equations* panel, you will see an equation of the form

$$f\left(u,u_t,u_{tt},t\right)=0, u\left(t_0\right)=u_0, u_t\left(t_0\right)=u_{t0} \tag{2.8}$$

- To convert Equations (2.4)–(2.7) to the desired form of Equation (2.8), enter the values as given in the following table.

| Name | $f(u, u_t, u_{tt}, t)$ | Initial value $(u\_0)$ | Initial value $(u\_t0)$ | Description |
|------|------------------------|------------------------|-------------------------|-------------|
| CA | u*CAt+k1*CA | 1 | 0 | Concentration of $A$ |
| CB | u*CBt−k1*CA+ k2*CB | 0 | 0 | Concentration of $B$ |
| CC | u*CCt − k2*CB | 0 | 0 | Concentration of $C$ |

This indicates that variables $CA, CB$, and $CC$ have initial values of 1, 0, and 0, respectively, at $z = 0$. Here, time dimension serves as the length dimension. In COMSOL, notation $u\dfrac{dC_A}{dz} + k_1C_A$ can be written as $u*CAt + k1*CA$, $u\dfrac{dC_B}{dz} - k_1C_A + k_2C_B$ can be written as $u*CBt - k1*CA + k2*CB$ and $u\dfrac{dC_C}{dz} - k_2C_B$ can be written as $u*CCt - k2*CB$.

This adjustment will convert Equations (2.4)–(2.7) to the desired form of Equation (2.8).

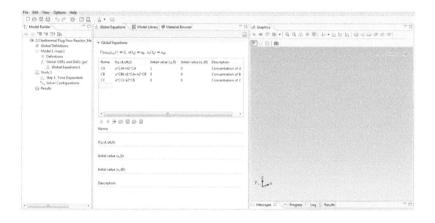

- In the *Model Builder*, select *Variables* by right clicking on the *Definitions* tab. Another window will open namely *Variables 1*. Click on it and define $u$: 1, $k1$: 1, and $k2$: 0.1.

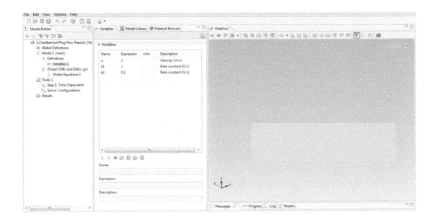

*Step 3*

- Now, expand the *Study 1* tab in *Model Builder*. Select *Step 1: Time Dependent*. Another window will open namely *Time Dependent*. Expand the *Study Settings* tab and set the range *Times: range (0,0.1,5)*.

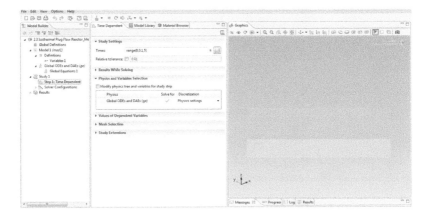

- Click on the *Compute (=)* button. A plot of *Concentrations versus Length* appears as shown in Figure 2.2.
- Save the simulation.

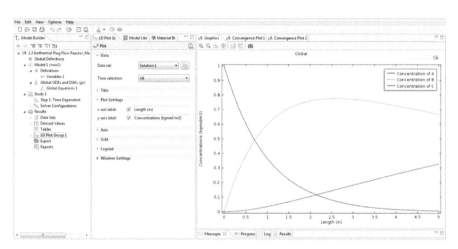

**FIGURE 2.2**   Concentrations as a function of length of the reactor [solution to Equations (2.4)–(2.7)].

### 2.3.3   *SIMULATION APPROACH (METHOD 2)*

*Step 1*

- Open COMSOL Multiphysics
- Select *0D Space Dimension* from the list of options. Hit the *next* arrow at the upper right corner.

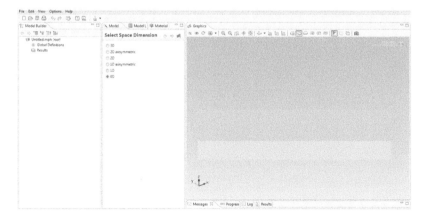

- Expand the *Chemical Species Transport* folder from the list of options in *Model Wizard* and select *Reaction Engineering*. Hit the *next* arrow again.

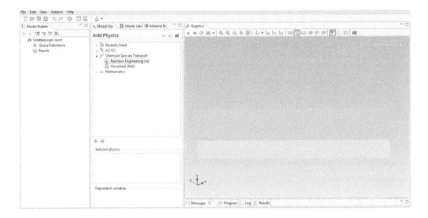

- Then select *Stationary Plug Flow* from the list of *Study Type* options, and click on the *Finish flag* at the upper right corner of the application.

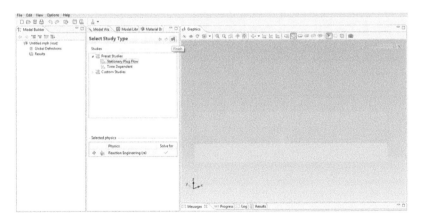

*Step 2*

- With the *Model Builder*, click on the *Reaction Engineering* tab and set the type of reactor and mixture as *Reactor type: Plug flow* and *Mixture: Liquid*.

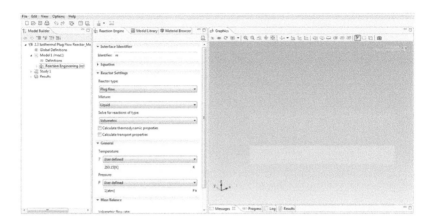

- Select *Reaction* option by right clicking on the *Reaction Engineering* tab and. Another window will open namely *Reaction 1*. Click on it.
- Select *Reaction 1* option available in the *Reaction Engineering* tab and in the formula edit window type, $A \Rightarrow B$. (Note: $\Rightarrow$ Equal to followed by greater sign).
- Select the following; *Reaction rate r: kf_1*c_A* and the *Rate constants* $k^f$: *1*.

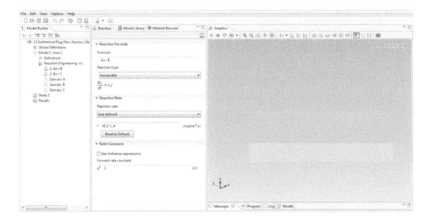

- Select the *Reaction* option by right clicking on the *Reaction Engineering* tab and. Another window will open namely *Reaction 2*. Click on it.
- Select *Reaction 2* option available in the *Reaction Engineering* tab and in the formula edit window type, $B \Rightarrow C$. (Note: $\Rightarrow$ Equal to followed by greater sign).

- Set the following; *Reaction rate r: kf_2\*c_B* and the *Rate constants kf: 0.1*.

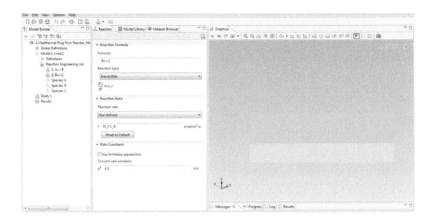

- With the *Reaction Engineering* tab, click on the *Species A* and set the *Rate expression R: -r_1* as $u\dfrac{dC_A}{dz} = -k_1 C_A = -r\_1$ and $u = 1$. Also, set the *Species Feed Stream* $F_0$: *1*.

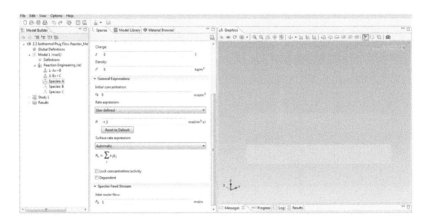

- With the *Reaction Engineering* tab, now click on the *Species B* and set the *Rate expression R: r_1-r_2* as $u\dfrac{dC_B}{dz} = k_1 C_A - k_2 C_B = r_1 - r\_2$ and $u = 1$. Also, set the *Species Feed Stream* $F_0$: *0*.

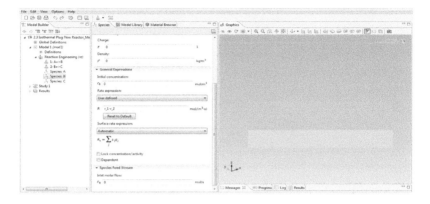

- With the *Reaction Engineering* tab, now click on the *Species C* and set the *Rate expression R: r_2* as $u\dfrac{dC_C}{dz} = k_2 C_B = r\_2$ and $u = 1$. Also, set the *Species Feed Stream $F_0$: 0*.

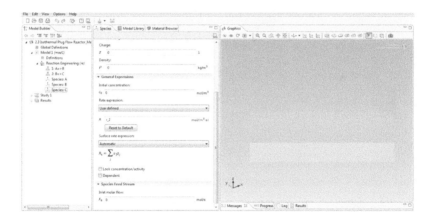

## Step 3

- Now, expand the *Study 1* tab in *Model Builder*. Select the *Step 1: Stationary Plug Flow*. Another window will open namely *Stationary Plug Flow*. Expand the *Study Settings* tab and set the range *Times: range (0,0.1,1)*.

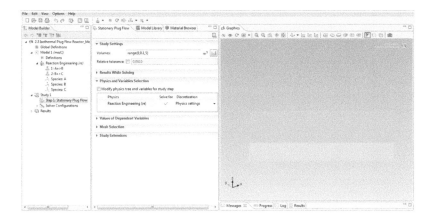

- Click on the *Compute (=)* button. A plot of *Concentrations versus Length* appears as shown in Figure 2.3.
- Save the simulation.

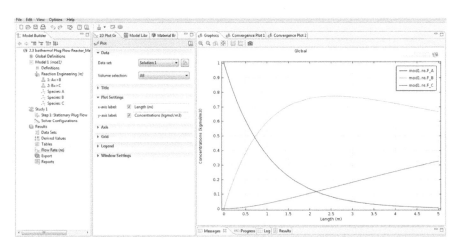

**FIGURE 2.3**    Concentrations as a function of length of the reactor [solution to Equations (2.4)–(2.7)].

## 2.4   SIMULATION OF THE BATCH REACTOR

### 2.4.1   PROBLEM STATEMENT

The following reactions take place in a batch reactor (Levenspiel, 2001):

$$A + B \xrightarrow{k_1} C \tag{2.9}$$

$$B + C \xrightarrow{k_2} D. \tag{2.10}$$

The differential equation for components $A$, $B$, $C$, and $D$ are

$$\frac{dC_A}{dt} = -k_1 C_A C_B \tag{2.11}$$

$$\frac{dC_B}{dt} = -k_1 C_A C_B - k_2 C_B C_C \tag{2.12}$$

$$\frac{dC_C}{dt} = k_1 C_A C_B - k_2 C_B C_C \tag{2.13}$$

$$\frac{dC_D}{dt} = k_2 C_B C_C. \tag{2.14}$$

The initial conditions are

$$\text{at } t = 0, C_A = 1\frac{\text{mol}}{\text{m}^3}, C_B(0) = 1\frac{\text{mol}}{\text{m}^3}, C_C(0) = 0\frac{\text{mol}}{\text{m}^3}, C_D = 0\frac{\text{mol}}{\text{m}^3}. \tag{2.15}$$

The rate constants are: $k_1 = 1$ m$^3$ / mol s and $k_1 = 0.1$ m$^3$ / mol s. Determine the concentration of species $A$, $B$, $C$, and $D$ at 10 s.

### 2.4.2   SIMULATION APPROACH

*Step 1*

- Select *0D Space Dimension* from the list of options by opening the COMSOL Multiphysics. Hit the *next* arrow at the upper right corner.

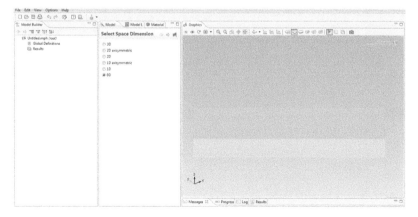

- Select and expand the *Mathematics* folder from the list of options in *Model Wizard*. Further choose *Global ODEs and DAEs* and hit *next tab*.

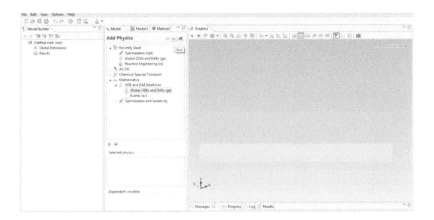

- Then select *Time Dependent* from the list of *Study Type* options, and click on the *Finish flag* at the upper right corner of the application.

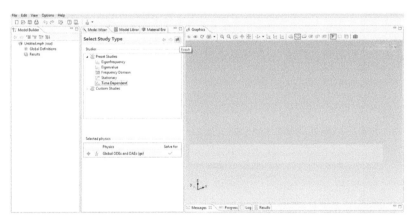

*Step 2*

- Now in the *Model Builder*, select *Global Equations* option. This will open a tab to enter the coefficients of the characteristics of a model equation.
- In the *Global Equations* panel, you will see an equation of the form

$$f\left(u,u_t,u_{tt},t\right)=0, u\left(t_0\right)=u_0, u_t\left(t_0\right)=u_{t0}. \tag{2.16}$$

To convert Equations (2.11)–(2.15) to the desired form of Equation (2.16), enter the values as given in the following table.

| Name | $f(u, u_t, u_{tt}, t)$ | Initial value ($u\_0$) | Initial value ($u\_t0$) | Description |
|------|------------------------|------------------------|-------------------------|-------------|
| CA | CAt+k1*CA*CB | 1 | 0 | Concentration of $A$ |
| CB | CBt+k1*CA*CB+k2 *CB*CC | 1 | 0 | Concentration of $B$ |
| CC | CCt–k1*CA*CB+k2 *CB*CC | 0 | 0 | Concentration of $C$ |
| CD | CDt–k2*CB*CC | 0 | 0 | Concentration of $D$ |

This indicates that variables $CA$, $CB$, $CC$, and $CD$ have initial values of 1, 1, 0, and 0, respectively, at $t=0$. In COMSOL notation, $\dfrac{dC_A}{dt}+k_1C_AC_B$ can be written as $CAt+k1*CA*CB$, $\dfrac{dC_B}{dt}+k_1C_AC_B+k_2C_BC_C$ can be written as $CBt+k1*CA*CB+k2*CB*CC$, $\dfrac{dC_C}{dt}-k_1C_AC_B+k_2C_BC_C$ can be written as $CCt-k1*CA*CB+k2*CB*CC$, and $\dfrac{dC_D}{dt}-k_2C_BC_C$ can be written as $CDt-k2*CB*CC$.

This adjustment will convert Equations (2.11)–(2.15) to the desired form of Equation (2.16).

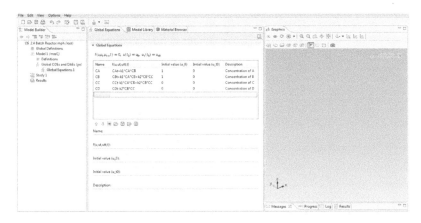

- In the *Model Builder*, select *Variables* by right clicking on the *Definitions* tab. Another window will open namely *Variables 1*. Click on it and define *u:* 1, *k*1: 1, and *k*2: 0.1.

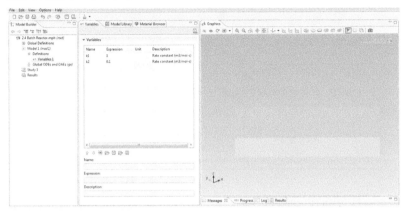

*Step 3*

- Now, expand the *Study 1* tab in *Model Builder*. Select the *Step 1: Time Dependent*. Another window will open namely *Time Dependent*. Expand the *Study Settings* tab and set the range *Times: range (0,0.1,10)*.

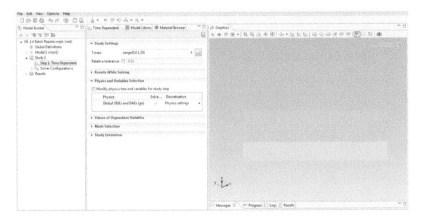

- Click on the *Compute (=)* button. A plot of *Concentrations versus Time* appears as shown in Figure 2.4.
- Save the simulation.

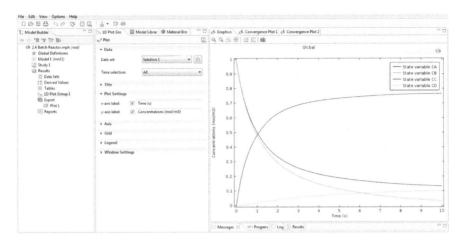

**FIGURE 2.4** Concentrations as a function of time [solution to Equations (2.11)–(2.15)].

## 2.5 SIMULATION OF CONTINUOUS STIRRED TANK REACTOR (CSTR)

### 2.5.1 PROBLEM STATEMENT

The energy equation for a stirred tank with coil heater (Ahuja, 2010) shown in Figure 2.5 is given by

$$MC_p \frac{dT}{dt} = m'C_p(T_1 - T) + Q'. \tag{2.17}$$

The heat transfer between the steam in the coil and the fluid is given by $Q = UA\Delta T_{\text{lmtd}} = UA(T_s - T)$. $\Delta T_{\text{lmtd}} = (T_s - T)$. The stirred tank heater is a square tank of 0.5 m on its sides and 2-m height. The water at 20 °C is flowing through the tank at 1 L/s. Determine the temperature of the tank after 3000 s,

where

$T_1$ = inlet stream temperature;
$T$ = outlet stream temperature (water temperature in the tank);
$T_s$ = coil/steam temperature (250 °C);
$U$ = overall heat transfer coefficient (200 W/m² K);
$A$ = outside area of the coil (1 m²);
$C_p$ = 4184 J/kg K.

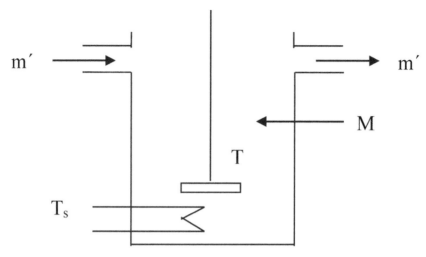

**FIGURE 2.5**   Stirred tank coil heater.

### 2.5.2   *SIMULATION APPROACH*

*Step 1*

- Select *0D Space Dimension* from the list of options by opening the COMSOL Multiphysics. Hit the *next* arrow at the upper right corner.

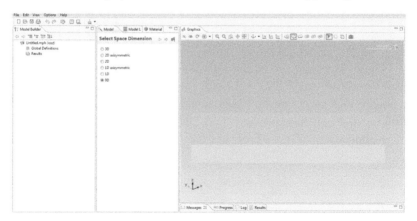

- Select and expand the *Mathematics* folder from the list of options in *Model Wizard*. Further choose *Global ODEs and DAEs* and hit *next tab*.

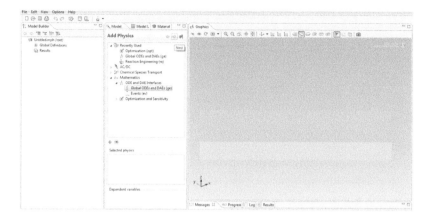

- Then select *Time Dependent* from the list of *Study Type* options, and click on the *Finish flag* at the upper right corner of the application.

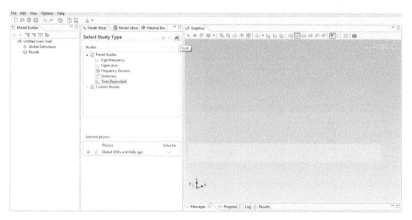

*Step 2*

- Now in the *Model Builder*, select *Global Equations* option. This will open a tab to enter the coefficients of the characteristics of a model equation.
- In the *Global Equations* panel, you will see an equation of the form

$$f(u, u_t, u_{tt}, t) = 0, u(t_0) = u_0, u_t(t_0) = u_{t0}. \tag{2.18}$$

To convert Equation (2.17) to the desired form of Equation (2.18), enter the values as given in the following table.

| Name | $f(u, u_t, u_{tt}, t)$ | Initial value ($u\_0$) | Initial value ($u\_t0$) | Description |
|------|------------------------|------------------------|--------------------------|-------------|
| $T$ | $M*Cp*Tt-m*Cp*(T1-T)-Q$  20 | 0 | | Temperature |

This indicates that variable $T$ has an initial value of 20 at $t = 0$. In COMSOL notation, $MC_p \dfrac{dT}{dt} - m'C_p\left(T_1 - T\right) - Q'$ can be written as $M*Cp*Tt-m*Cp*(T1-T)-Q$.

This adjustment will convert Equation (2.17) to the desired form of Equation (2.18).

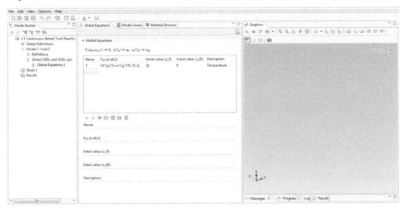

- In the *Model Builder*, select *Variables* by right clicking on the *Definitions* tab. Another window will open namely *Variables 1*. Click on it and define *M: 500, m: 1, $C_p$: 4184, T1: 20, Ts: 250, UA: 200,* and *Q: UA(Ts – T)*.

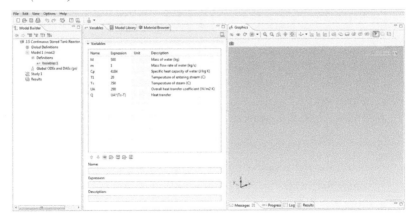

*Step 3*

- Now, expand the *Study 1* tab in *Model Builder*. Select the *Step 1: Time Dependent*. Another window will open namely *Time Dependent*. Expand the *Study Settings* tab and set the range *Times: range (0,10,3000)*.

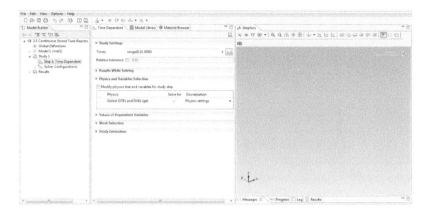

- Click on the *Compute (=)* button. A plot of *Temperature versus Time* appears as shown in Figure 2.6.
- Save the simulation.

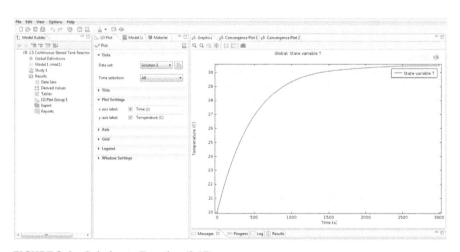

**FIGURE 2.6**   Solution to Equation (2.17).

## 2.6  SIMULATION OF NONLINEAR EQUATIONS WITH FUEL CELL APPLICATIONS

### 2.6.1  PROBLEM STATEMENT

Consider the following reaction to make hydrogen for fuel cell (Finlayson, 1980):

$$CO + H_2O \leftrightarrow CO_2 + H_2. \tag{2.19}$$

The equilibrium composition is given by

$$148.4 = \frac{yCO_2\,yH_2}{yCOH_2O} \tag{2.20}$$

Table 2.1 represents the mole balance.

**TABLE 2.1**   Equilibrium of Water–Gas Shift Reaction

| Species | Start | End | $y_i$ |
|---------|-------|-----|-------|
| CO | 1 | $1 - x$ | $(1 - x)/2$ |
| $H_2O$ | 1 | $1 - x$ | $(1 - x)/2$ |
| $CO_2$ | | $x$ | $x/2$ |
| $H_2$ | | $x$ | $x/2$ |
| Total | 2 | 2 | $1$ |

Put the mole fractions into Equation (2.20) and simplify

$$148.4 = \frac{(x/2)(x/2)}{[(1-x)/2][(1-x)/2]} = \frac{x^2}{(1-x)^2}. \tag{2.21}$$

Solve the nonlinear equation (2.21).

### 2.6.2  SIMULATION APPROACH

*Step 1*

- Select *0D Space Dimension* from the list of options by opening the COMSOL Multiphysics. Hit the *next* arrow at the upper right corner.

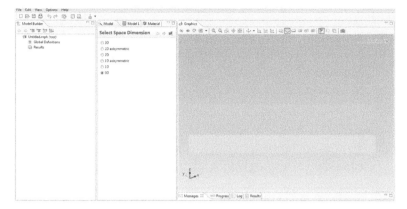

- Select and expand the *Mathematics* folder from the list of options in *Model Wizard*. Further choose *Global ODEs and DAEs* and hit *next tab*.

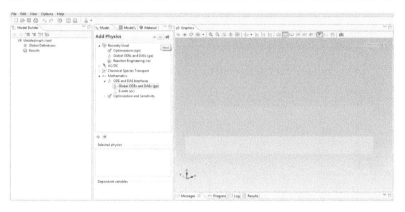

- Then select *Stationary* from the list of *Study Type* options, and click on the *Finish flag* at the upper right corner of the application.

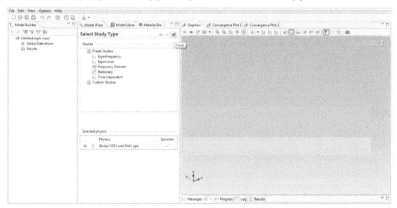

*Step 2*

- Now in the *Model Builder*, select *Global Equations* option. This will open a tab to enter the coefficients of the characteristics of a model equation.
- In the *Global Equations* panel, you will see an equation of the form

$$f\left(u, u_t, u_{tt}, t\right) = 0, u\left(t_0\right) = u_0, u_t\left(t_0\right) = u_{t0}. \tag{2.22}$$

To convert Equation (2.21) to the desired form of Equation (2.22), enter the values as given in the following table.

| Name | $f(u, u_t, u_{tt}, t)$ | Initial value $(u_0)$ | Initial value $(u_{t0})$ | Description |
|------|------------------------|------------------------|---------------------------|-------------|
| $x$ | $148.4 - (x^2)/(1 - x)^2$ | 0.5 | 0 | |

This indicates that variable $x$ has an initial guess of 0.5 at $t = 0$. In COMSOL notation $148.4 - \dfrac{x^2}{(1-x)^2}$ can be written as $148.4 - (x^2)/(1-x)^2$. This adjustment will convert Equation (2.21) to the desired form of Equation (2.22).

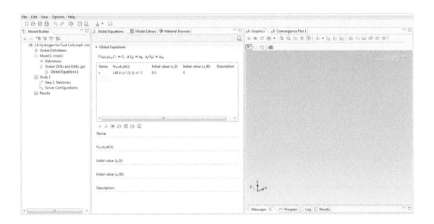

*Step 3*

- Now go to *Study* option in the model pellet tab. Click on the *Compute (=)* button. The error reduces from 0.97 to 5.8e-5 in six iterations.

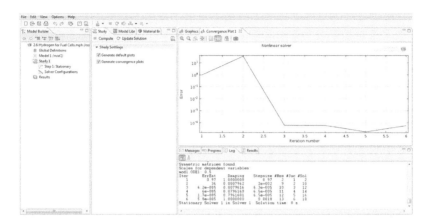

## Step 4

- To calculate the variable *x*, select *Derived Values* option available in the *Results* tab.
- Now, select the *Global Evaluation* option by right clicking on the *Derived Values* tab. Another window will open namely *Global Evaluation 1*. Click on it.
- Click on "= Evaluate" button at the top. The value of variable *x* is 1.08943. But, *x* cannot be more than 1 as it represents the mole of CO and the total available is 1.0.

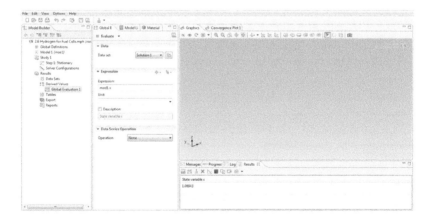

- Now in the *Model Builder*, select *Global Equations 1* option and change the initial guess from 0.5 to 0.9.

- Now go again to *Study* option in the model pellet tab. Click on the *Compute (=)* button.
- Now, again select the *Global Evaluation 1* option by right clicking on the *Derived Values* tab.
- Click on "= Evaluate" button at the top. The value of variable $x$ is 0.92414. It indicates that most nonlinear algebraic equations have more than one solution.

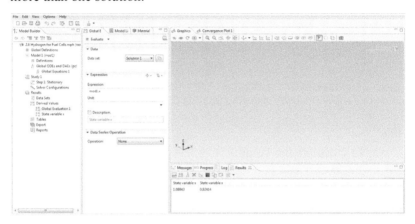

## 2.7 SIMULATION OF CHEMICAL REACTORS WITH MASS TRANSFER LIMITATIONS

### 2.7.1 PROBLEM STATEMENT

The differential equation for components $A$, $B$, and $C$ are given as

$$u\frac{dC_A}{dz} = -2kC_{A,s}^2, u\frac{dC_B}{dz} = +kC_{A,s}^2, u\frac{dC_C}{dz} = 0. \qquad (2.23)$$

$C_s$ is the concentration on the surface, expressed as the kmol of a species per volume of the catalyst.

Equation (2.24) relates the rate of mass transfer to the catalyst to the rate of reaction (Finlayson, 2006)

$$k_m a\left(C_A - C_{A,s}\right) = kC_{A,s}^2 \qquad (2.24)$$

where $k_m$, $a$, and $k$ are mass transfer coefficient (m/s), surface area exposed per volume of the reactor (m²/m³), and rate of reaction rate constant (m³/kmol s), respectively. This problem combines ordinary differential equations with nonlinear algebraic equations.

Use the following initial conditions:

$$C_A(0) = 2\,\text{kmol}/\text{m}^3, C_B(0) = 0, C_C(0) = 2\,\text{kmol}/\text{m}^3 \qquad (2.25)$$

and we take $u = 0.5$ m/s, $k = 0.3$ m$^3$/kmol s, $k_m a = 0.2$, and the total reactor length as $z = 0.5$ m.

## 2.7.2  SIMULATION APPROACH

*Step 1*

- Select *0D Space Dimension* from the list of options by opening the COMSOL Multiphysics. Hit the *next* arrow at the upper right corner.

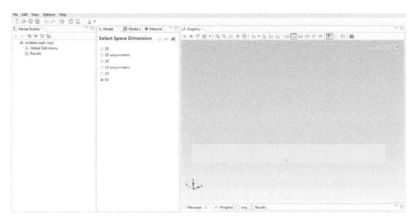

- Select and expand the *Mathematics* folder from the list of options in *Model Wizard*. Further choose *Global ODEs and DAEs* and hit *next tab*.

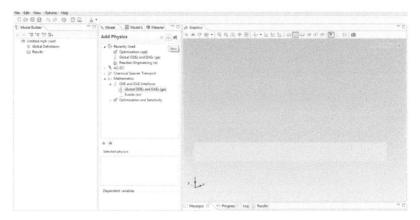

- Then select *Time Dependent* from the list of *Study Type* options, and click on the *Finish flag* at the upper right corner of the application.

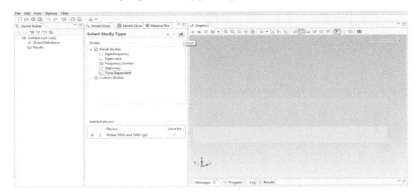

*Step 2*

- Now in the *Model Builder*, select *Global Equations* option. This will open a tab to enter the coefficients of the characteristics of a model equation.
- In the *Global Equations* panel, you will see an equation of the form

$$f\left(u, u_t, u_{tt}, t\right) = 0, u\left(t_0\right) = u_0, u_t\left(t_0\right) = u_{t0}. \tag{2.26}$$

To convert Equations (2.23)–(2.25) to the desired form of Equation (2.26), enter the values as given in the following table.

| Name $f(u, u_t, u_{tt}, t)$ | | Initial value ($u_0$) | Initial value ($u_{t0}$) | Description |
|---|---|---|---|---|
| CA | $u*Cat + 2*k*CAs^2$ | 2 | 0 | Concentration of $A$ |
| CB | $u*CBt - k*CAs^2$ | 0 | 0 | Concentration of $B$ |
| CC | $u*CCt$ | 2 | 0 | Concentration of $C$ |
| CAs | $kma*CA - kma*CAs - k*CAs^2$ 0.5 | | 0 | Concentration of $A$ on the surface |

This indicates that variable *CA* has an initial value of two at $z = 0$, variable *CB* has an initial value of zero at $z = 0$, variable *CC* has an initial value of two at $z = 0$, and variable *CA* has an initial guess of 0.5 at $z = 0$. Here, time dimension serves as the length dimension. In COMSOL notation, $u\dfrac{dC_A}{dz} + 2kC_{A,s}^2$ can be written as $u*CAt+2*k*CAs^2$, $u\dfrac{dC_B}{dz} - kC_{A,s}^2$ can be written as $u*CBt - k*CAs^2$,

$u\dfrac{dC_C}{dz}$ can be written as $u*CCt$ and $k_m a\left(C_A - C_{A,s}\right) - kC_{A,s}^2$ can be written as $kma*CA - kma*CAs - k*CAs^2$.

This adjustment will convert Equations (2.23)–(2.25) to the desired form of Equation (2.26).

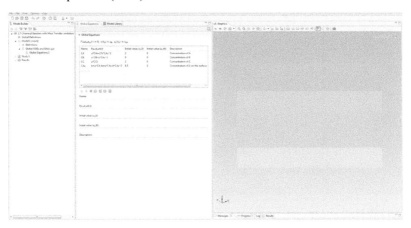

- In the *Model Builder*, select *Variables* by right clicking on the *Definitions* tab. Another window will open namely *Variables 1*. Click on it and define *u: 0.5, k: 0.3,* and *kma: 0.2.*

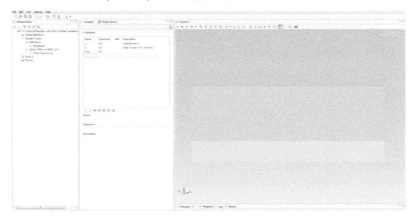

*Step 3*

- Now, expand the *Study 1* tab in *Model Builder*. Select the *Step 1: Time Dependent*. Another window will open namely *Time Dependent*. Expand the *Study Settings* tab and set the range *Times: range (0,0.1,2.5).*

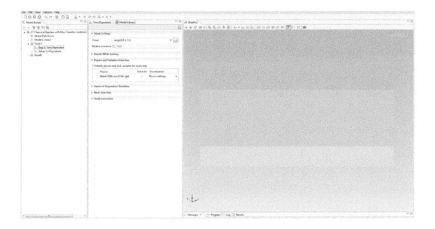

- Click on the *Compute (=)* button. A plot of *Concentrations versus Length* appears as shown in Figure 2.7.
- Save the simulation.

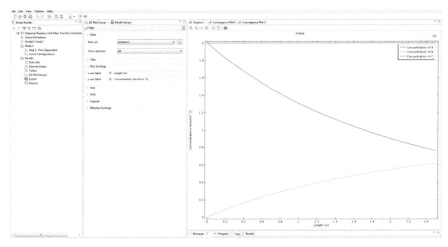

**FIGURE 2.7**   Concentration as a function of length [solution to Equations (2.23)–(2.25)].

## 2.8   SIMULATION OF RECTANGULAR FIN

### 2.8.1   *PROBLEM STATEMENT*

Consider the heat conduction in fin as shown in Figure 2.8 (Ghoshdastidar, 1998). The energy equation for the fin at the steady state is

$$\frac{d^2\theta}{dX^2} - (mL)^2\theta = 0 \tag{2.27}$$

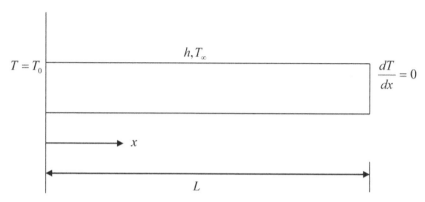

**FIGURE 2.8** Physical domain of the rectangular fin.

The above partial differential equation (2.27) has been discretized by using the central difference scheme and the following matrix has been obtained.

$$\begin{bmatrix} 3.0 & -1.2 & & \\ -1.2 & 3.0 & -1.2 & \\ & -1.2 & 3.0 & -1.2 \\ & & -2.4 & 3.0 \end{bmatrix}\begin{bmatrix} \theta_1 \\ \theta_2 \\ \theta_3 \\ \theta_4 \end{bmatrix} = \begin{bmatrix} 1 \\ 0 \\ 0 \\ 0 \end{bmatrix}$$

The above matrix can be rearranged in the form of following sets of linear equations:

$$3\theta_1 - 1.2\theta_2 + 0\theta_3 + 0\theta_4 = 1 \tag{2.28a}$$
$$-1.2\theta_1 + 3\theta_2 - 1.2\theta_3 + 0\theta_4 = 0 \tag{2.28b}$$
$$0\theta_1 - 12\theta_2 + 3\theta_3 - 12\theta_4 = 0 \tag{2.28c}$$
$$0\theta_1 + 0\theta_2 - 2.4\theta_3 - 3\theta_4 = 0. \tag{2.28d}$$

Find the values of dimensionless temperature with COMSOL Multiphysics.

### 2.8.2 SIMULATION APPROACH

*Step 1*

- Select *0D Space Dimension* from the list of options by opening the COMSOL Multiphysics. Hit the *next* arrow at the upper right corner.

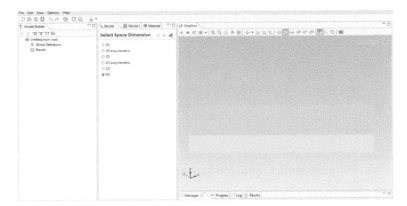

- Select and expand the *Mathematics* folder from the list of options in *Model Wizard*. Further choose *Global ODEs and DAEs* and hit *next tab*.

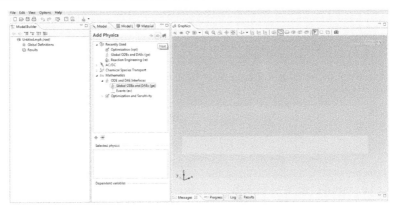

- Then select *Stationary* from the list of *Study Type* options, and click on the *Finish flag* at the upper right corner of the application.

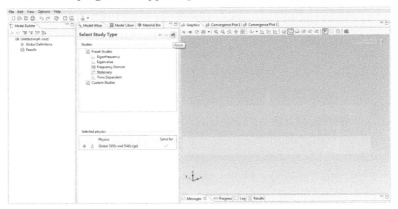

*Step 2*

- Now in the *Model Builder*, select *Global Equations* option. This will open a tab to enter the coefficients of the characteristics of a model equation.
- In the *Global Equations* panel, you will see an equation of the form

$$f\left(u, u_t, u_{tt}, t\right) = 0, u\left(t_0\right) = u_0, u_t\left(t_0\right) = u_{t0}. \tag{2.29}$$

To convert Equations (2.28a)–(2.28d) to the desired form of Equation (2.29), enter the values as given in the following table.

| Name | $f(u, u_t, u_{tt}, t)$ | Initial value ($u_0$) | Initial value ($u_{t0}$) | Description |
|------|------------------------|------------------------|---------------------------|-------------|
| $u1$ | $3 * u1 - 1.2 * u2 + 0 * u3 + 0 * u4 - 1$ | 0 | 0 | Variable to represent $\theta_1$ |
| $u2$ | $-1.2 * u1 + 3 * u2 - 1.2 * u3 + 0 * u4$ | 0 | 0 | Variable to represent $\theta_2$ |
| $u3$ | $0 * u1 - 1.2 * u2 + 3 * u3 - 1.2 * u4$ | 0 | 0 | Variable to represent $\theta_3$ |
| $u4$ | $0 * u1 + 0 * u2 - 2.4 * u3 + 3 * u4$ | 0 | 0 | Variable to represent $\theta_4$ |

This indicates that variables $u1$, $u2$, $u3$, and $u4$ are used to represent variables $\theta_1$, $\theta_2$, $\theta_3$, and $\theta_4$ respectively. In COMSOL notation, $3\theta_1 - 1.2\theta_2 + 0\theta_3 + 0\theta_4 - 1$ can be written as $3 * u1 - 1.2 * u2 + 0 * u3 + 0 * u4 - 1$, $-1.2\theta_1 + 3\theta_2 - 1.2\theta_3 + 0\theta_4$ can be written as $-1.2 * u1 + 3 * u2 - 1.2 * u3 + 0 * u4$, $0\theta_1 - 1.2\theta_2 + 3\theta_3 - 1.2\theta_4$ can be written as $0 * u1 - 1.2 * u2 + 3 * u3 - 1.2 * u4$, and $0\theta_1 + 0\theta_2 - 2.4\theta_3 - 3\theta_4$ can be written as $0 * u1 + 0 * u2 - 2.4 * u3 + 3 * u4$.

This adjustment will convert Equations (2.28a)–(2.28d) to the desired form of Equation (2.29).

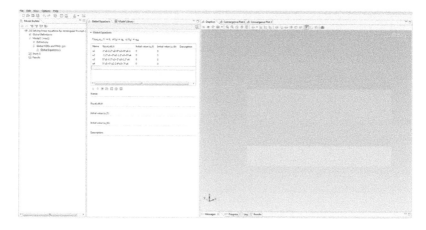

*Step 3*

- Click on the *Compute* (=) button. Save the simulation.
- To calculate the variable $u1$, select *Derived Values* option available in the *Results* tab.
- Now, click on "= Evaluate" button at the top. The value of variable $u1$ is 0.42153.
- Similarly, "Evaluate" the "State variables" $u2$, $u3$, and $u4$. The following solution is obtained:
  $u2 = 0.22049$, $u3 = 0.1297$, and $u4 = 0.10376$.

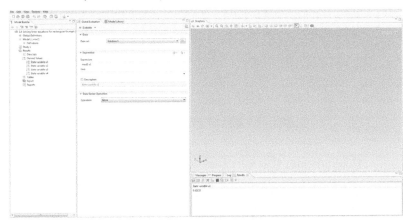

## 2.9   SIMULATION OF THREE REACTORS LINKED BY PIPES

### 2.9.1   PROBLEM STATEMENT

Find the concentrations of three reactors linked by pipes shown in Figure 2.9.

MASS BALANCES:

First Reactor:

$$Q21c2 + 400 = Q13c1 + Q12c1$$
$$20c2 + 400 = 40c1 + 80c1$$
$$120c1 - 20c2 = 400. \qquad (2.30a)$$

Second Reactor:

$$Q12c1 = Q21c2 + Q23c2$$
$$80c1 = 20c2 + 60c2$$
$$80c1 - 80c2 = 0$$
$$c1 - c2 = 0. \qquad (2.30b)$$

Third Reactor:

$$200 + Q23c2 + Q13c1 = Q33c3$$
$$200 + 60c2 + 40c1 = 120c3$$
$$40c1 + 60c2 - 120c3 = -200. \qquad (2.30c)$$

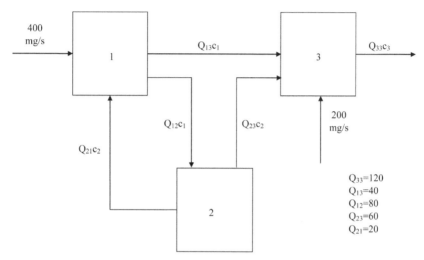

**FIGURE 2.9**   Three reactors linked by pipes.

## 2.9.2   *SIMULATION APPROACH*

*Step 1*

- Select *0D Space Dimension* from the list of options by opening the COMSOL Multiphysics. Hit the *next* arrow at the upper right corner.

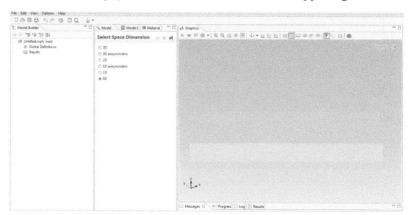

- Select and expand the *Mathematics* folder from the list of options in *Model Wizard*. Further choose *Global ODEs and DAEs* and hit *next tab*.

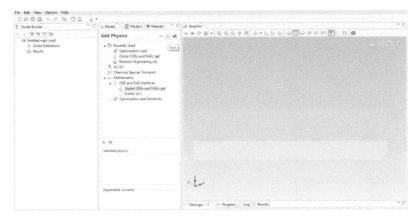

- Then select *Stationary* from the list of *Study Type* options, and click on the *Finish flag* at the upper right corner of the application.

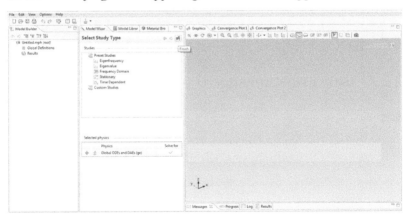

## Step 2

- Now in the *Model Builder*, select *Global Equations* option. This will open a tab to enter the coefficients of the characteristics of a model equation.
- In the *Global Equations* panel, you will see an equation of the form

$$f\left(u, u_t, u_{tt}, t\right) = 0, u\left(t_0\right) = u_0, u_t\left(t_0\right) = u_{t0}. \qquad (2.31)$$

To convert Equations (2.30a)–(2.30c) to the desired form of Equation (2.31), enter the values as given in the following table.

| Name $f(u, u_t, u_{tt}, t)$ | | Initial value $(u_0)$ | Initial value $(u_{t0})$ | Description |
|---|---|---|---|---|
| $c1$ | $120*c1 - 20*c2 - 400$ | 0 | 0 | Concentration of reactor 1 |
| $c2$ | $c1 - c2$ | 0 | 0 | Concentration of reactor 2 |
| $c3$ | $40*c1 + 60*c2 - 120*c3 + 200$ | 0 | 0 | Concentration of reactor 3 |

In COMSOL notation, $120c1 - 20c2 - 400$ can be written as $120*c1 - 20*c2 - 400$, $c1 - c2$ can be written as $c1-c2$, and $40c1 + 60c2 - 120c3 + 200$ can be written as $40*c1 + 60*c2 - 120*c3 + 200$.

This adjustment will convert Equations (2.30a)–(2.30c) to the desired form of Equation (2.31).

### Step 3

- Click on the *Compute (=)* button. Save the simulation.
- To calculate the variable c1, select *Derived Values* option available in the *Results* tab.
- Now, click on "= Evaluate" button at the top. The value of variable $c1$ is 4.
- Similarly, "Evaluate" the "State variables" $c2$ and $c3$. The following solution is obtained:
  $c2 = 4$ and $c3 = 5$.

## 2.10   PROBLEMS

1.  The equations for components $A$, $B$, and $C$ in the plug flow reactor are given as

$$u\frac{dC_A}{dz} = -2kC_A^2, u\frac{dC_B}{dz} = +kC_A^2, u\frac{dC_C}{dz} = 0$$

where $u$ is the velocity and $z$ is the distance from the inlet.

At the inlet, $C_A(0) = 2$ kmol/m³, $C_B(0) = 0$, and $C_C(0) = 2$ kmol/³, and we take $u = 0.5$ m/s, $k = 0.3$ m³/kmol s, and reactor length as $z = 2.5$ m.

2. Consider the reactor oxidizing $SO_2$ to form $SO_3$. The equations are

$$\frac{dX}{dz} = -50R'$$

$$\frac{dT}{dz} = -4.1(T - T_{surr}) + 1.02 \times 10^4 R'$$

where the reaction rate is

$$R' = \frac{X[1 - 0.167(1-X)]^{1/2} - 2.2(1-X)/K_{eq}}{[k_1 + k_2(1-X)]^2}$$

$\ln k_1 = -14.96 + 11070/T$; $\ln k_2 = -1.331 + 2331/T$; $K_{eq} = -11.02 + 11570/T$.

The parameters are $T_{surr} = 673.2$, $T(0) = 673.2$, and $X(0) = 1$.
$X$ = concentration of $SO_2$/inlet concentration;
$1-X$ = fractional conversion;
$T$ = temperature (K).

3. The following reaction takes place in a batch reactor:

$$A \rightarrow B$$

The equation for component $A$ is

$$\frac{dC_A}{dt} = -kC_A.$$

Initial condition: at $t = 0$, $C_A = 1$ mol/m³.
Rate constant = 1 s⁻¹.
Compute the concentration of $A$ at 5 s considering the step size of 0.1 s.

4. The following reaction takes place in a plug flow reactor:

$$A \rightarrow B.$$

The equation for component $A$ is

$$u\frac{dC_A}{dz} = -kC_A$$

Initial condition: at $z = 0$, $C_A = 1$ mol/m³.
Rate constant = 0.1 s⁻¹.
Compute the concentration of $A$ at 5 m considering the step size of 0.01 m. Take $u = 1$ m/s.

5. The reaction $A \rightarrow B$ takes in a stirred vessel reactor having same inlet and outlet feed rate of $F = 1$ L/s. The vessel may be considered perfectly mixed. The rate of consumption of $A$ equals $kC_A$, where $k = 1$ s$^{-1}$. Compute the concentration of $A$ at 5 s considering the step size of 0.01 s. Take $C_A = 1 \dfrac{mol}{m^3}$ and $V = 10$ L.

6. The following reaction takes place in a batch reactor:

$$A \xrightarrow{k_1} B \xrightarrow{k_2} C.$$

Equations for components $A$, $B$, and $C$ are

$$\frac{dC_A}{dt} = -k_1 C_A$$

$$\frac{dC_B}{dt} = k_1 C_A - k_2 C_B$$

$$\frac{dC_C}{dt} = k_2 C_B.$$

Initial conditions:

$t = 0$, $C_A = 1$ mol/m$^3$, $C_B = 0$ mol/m$^3$, and $C_C = 0$ mol/m$^3$.

Take $k_1 = 1$ s$^{-1}$ and $k_2 = 1$ s$^{-1}$.

Compute the concentration of $A$, $B$, and $C$ at 2 s considering the step size of 0.1 s.

7. The stirred vessel contains 800 kg of solvent at 30 °C having the same inlet and exit solvent rate of 12 kg/min. The heat transfer between the steam in the coil and the fluid is given by $Q = UA(T_s - T)$. Determine the temperature of the tank after 200 s.

where

$T_s$ = coil/steam temperature (200 °C);
$UA$ = overall heat transfer coefficient (15 kJ/min K);
$C_p$ = 2.3 kJ/kg K.

8. The flowsheet for recovery of acetone is shown in Figure 2.10. Find the concentration of acetone in the vapor stream and flow rates of product streams from the information given in the matrix.

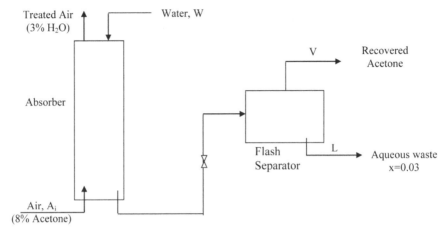

**FIGURE 2.10**   Recovery of acetone.

$$
\begin{bmatrix}
0.97 & 0 & 0 \\
0 & 0.03 & 0.615 \\
0.03 & 0.385 & 0.97
\end{bmatrix}
\begin{bmatrix}
A_0 \\
L \\
V
\end{bmatrix}
=
\begin{bmatrix}
0.92 \times 600 \\
0.08 \times 600 \\
500
\end{bmatrix}
$$

9.  The position of three ball are hanging vertically by springs are modeled by the following steady-state force balances (Chapra and Canale, 1998):

$$k(x_2 - x_1) + m_1 g - kx_1 = 0$$
$$k(x_3 - x_2) + m_2 g - k(x_2 - x_1) = 0$$
$$m_3 g - k(x_3 - x_2) = 0$$

If $g = 9.81$ m/s$^2$, $m_1 = 2.5$ kg, $m_2 = 3.5$ kg, $m_3 = 3$ kg, and $k = 12$ N/m, solve for displacements $x$.

10. The air at 25 °C and 1 atm having density and viscosity of 1.23 kg/m$^3$ and $1.79 \times 10^{-5}$ kg/m×s, respectively, is flowing at the velocity of 60 m/s through a 6-mm-diameter tube having roughness $\varepsilon = 0.0018$ mm. Calculate the friction factor and pressure drop in a 2-m tube using the following equations:

$$\frac{1}{\sqrt{f}} = -2.0\log\left(\frac{\varepsilon / D}{3.7} + \frac{2.51}{Re\sqrt{f}}\right)$$

$$\Delta P = \frac{fLV^2\rho}{2D}$$

11. The three distillation columns shown in Figure 2.11 are used to separate benzene (1), toluene (2), styrene (3), and xylene (4). Determine the compositions of streams *B* and *D* and molar flow rates of streams *D1, B1, D2, B2, B,* and *D.*

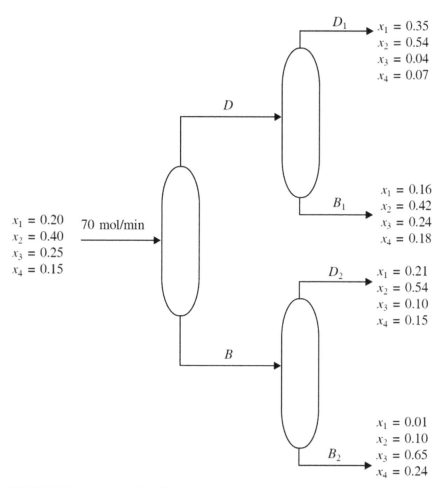

$D_1$   $x_1 = 0.35$
$x_2 = 0.54$
$x_3 = 0.04$
$x_4 = 0.07$

$D$

$x_1 = 0.20$   70 mol/min
$x_2 = 0.40$
$x_3 = 0.25$
$x_4 = 0.15$

$B_1$   $x_1 = 0.16$
$x_2 = 0.42$
$x_3 = 0.24$
$x_4 = 0.18$

$D_2$   $x_1 = 0.21$
$x_2 = 0.54$
$x_3 = 0.10$
$x_4 = 0.15$

$B$

$B_2$   $x_1 = 0.01$
$x_2 = 0.10$
$x_3 = 0.65$
$x_4 = 0.24$

**FIGURE 2.11** Sequence of distillation columns.

12. A component material balance around a reactor yields the steady-state equation

$$0 = \frac{F}{V}C_{in} - \frac{F}{V}C - kC^3$$

where $\dfrac{F}{V} = 0.1$ min⁻¹, $C_{in} = 1$ kmol/m³, and $k = 0.05$ m⁶/kmol² min.

Find the root of the above equation with initial guess of 1.0.

13. Taking guess values of $x = 0.1$ and $y = 0.5$, obtain the values of $x$ and $y$

$f(x, y) = e^{2x} + xy - 1 = 0$

$g(x, y) = \cos xy + x + y - 1 = 0.$

14. The first-order exothermic reaction takes place in a CSTR. The steady-state analysis of the non-isothermal CSTR yields equations as

$$Y_{ss} = \frac{BX_{ss}}{1 + \beta}$$

$$F(X_{ss}) = -(1 + \beta)Y_{ss} + BD_a e^{Y_{ss}}(1 - X_{ss}) = 0.$$

In the above equations, $X_{ss}$ and $Y_{ss}$ are dimensionless concentration and temperature, respectively. Given the dimensionless parameter values $D_a = 0.082$, $B = 22.0$, and $\beta = 3.0$, find all the possible roots of $F(X_{ss}) = 0$.

15. If the surface area of cylinder is 5 m² and the ratio of height to diameter is 2, express the surface area in terms of the diameter and find the value of the diameter for this surface area.

16. Determine the molal volume $(V)$ of carbon dioxide and oxygen for different combinations of temperature and pressure to select the vessel based on the following data:
$R = 0.082054$ L atm/mol K.

**Carbon dioxide:**

$a = 3.489$

$b = 0.03987$

**Oxygen:**

$a = 1.462$

$b = 0.02961.$

The design pressures of interest are 2, 20, and 200 atm for temperature combinations of 400, 600, and 800 K.

# CHAPTER 3

# Transport Processes

## 3.1 INTRODUCTION

Transport phenomena include the study of heat, mass, and momentum transfer. In this chapter, the application of *COMSOL* is discussed to solve problems in transport processes in one dimension using *Mathematics* and *Chemical Engineering* modules. The examples discussed here include the calculation of Dirichlet and Neumann boundary conditions, steady-state heat transfer in the rectangular fin, the illustration on reaction–diffusion considering spherical domain, flow of Newtonian and non-Newtonian fluid in a pipe, unsteady-state heat transfer in a hot infinite plate, linear adsorption, and heat transfer in an infinitely long cylinder.

## 3.2 SIMULATION OF DIFFERENT TYPES OF BOUNDARY CONDITIONS

### 3.2.1 PROBLEM STATEMENT

Here, we will solve a 1D partial differential equation (PDE) using different types of boundary conditions. For that purpose, the following PDE and boundary conditions will be used:

$$\frac{d^2u}{dx^2} + \frac{du}{dx} + u = 2\cos(x) + x + 1, \qquad 0 \le x \le 2\pi \tag{3.1}$$

1. Dirichlet boundary condition: $u(0) = 0$, $u(2\pi) = 2\pi$. $\qquad\qquad$ (3.2)

2. Neumann boundary condition: $u(0) = 0$, $\left.\dfrac{du}{dx}\right|_{x=2\pi} = 3$. $\qquad$ (3.3)

### 3.2.2  SIMULATION APPROACH

*Step 1*

- Open COMSOL Multiphysics.
- Select *1D Space Dimension* from the list of options. Hit the *next* arrow at the upper right corner.

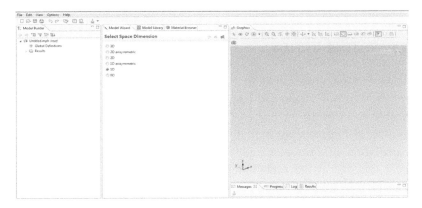

- Select and expand the *Mathematics* folder from the list of options in *Model Wizard*. Further select and expand *PDE Interface* and click on the *Coefficient Form PDE (c)*. Again hit the *next* arrow.

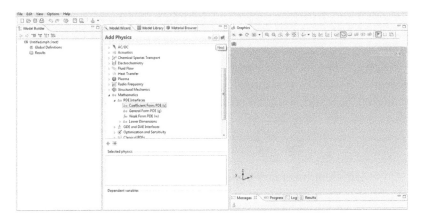

- Then select *Stationary* from the list of *Study Type* options, and click on the *Finish flag* at the upper right corner of the application.

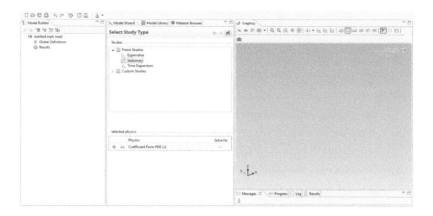

### Step 2

- With the *Model Builder*, right click on the *Geometry* option. From the list of options, click on the *Interval* taskbar.
- In the *Interval* environment, select *Intervals*1, *Left endpoint: 0 and Right endpoint: 2\*pi*. Click on the *Build All* option at the upper part of tool bar. A straight line graph will appear on the right side of the application window.

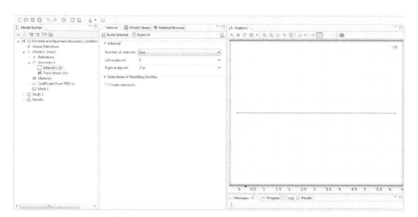

### Step 3

- Now, in the *Model Builder*, select *Coefficient Form PDE 1* option. This will open a tab to enter the coefficients of the characteristics of a model equation.
- In the *Domain Selection* panel, you will see an equation of the form

$$e_a \frac{\partial^2 u}{\partial t^2} + d_a \frac{\partial u}{\partial t} + \nabla \cdot \left( -c\nabla u - \alpha u + \gamma \right) + \beta \cdot \nabla u + au = f \qquad (3.4)$$

where $\nabla = \left[ \dfrac{\partial}{\partial x} \right]$.

In order to solve the governing differential equation, we need to assign the coefficients in above equation a suitable value.

To convert Equation (3.4) to the desired form of Equation (3.1), the value of coefficients in Equation (3.4) to be changed as follows:

$c = -1$      $\alpha = -1, \beta = 0, \gamma = 0$

$a = 1$      $f = 2*\cos(x) + x + 1$

$e_a = 0$      $d_a = 0.$

This adjustment will reduce Equation (3.4) to Equation (3.1) format.

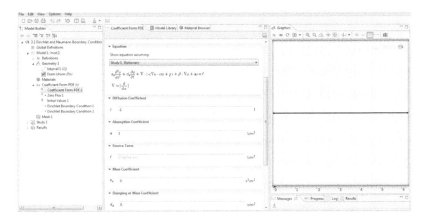

*Step 4*

- Now, right click on the *Coefficient Form PDE* tab in *Model Builder.* Select the *Dirichlet Boundary Condition* option. Another window will open namely *Dirichlet Boundary Condition 1.* Click on it.
- Select the left point on the horizontal line graph. Click on the "+" sign at the top right corner; this will add *boundary 1* in the *Boundary Selection* tab. At this point, put *r: 0.* This step will add a *Dirichlet boundary condition: u(0)=0* as given in Equation (3.2) (Note: as *u = r, r = 0, u = 0*).

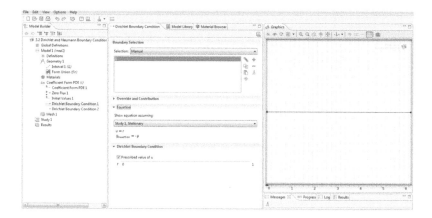

- Again right click on the *Coefficient Form PDE* tab in *Model Builder.* Select the *Dirichlet Boundary Condition* option again. Another window will open namely *Dirichlet Boundary Condition 2.* Click on it.
- Select the right point on the horizontal line graph. Click on the "+" sign at the top right corner, this will add *boundary 2* in the *Boundary Selection* tab. At this point, put *r: 2\*pi.* This step will add a *Dirichlet boundary condition:* $u(2\pi) = 2\pi$ as given in Equation (3.2) (Note: as *u = r, r = 2\*pi, u = 2\*pi*).

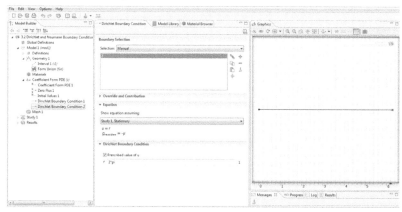

*Step 5*

- Click on the *Mesh* option in *Model Builder.* Select *Normal Mesh* Type. Click on *Build All* option at the top of ribbon. A dialogue box will appear in the *Message* tab as: "*Complete mesh consists of 15 elements.*"

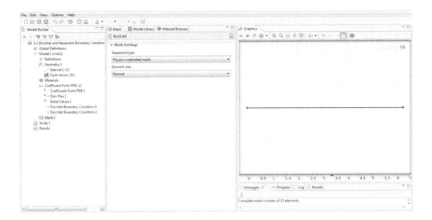

- Now, go to *Study* option in the model pellet tab. Click on the *Compute (=)* button. A graph will appear giving the profile of *u(x)* versus *x* as shown in Figure 3.1.
- Save the simulation.

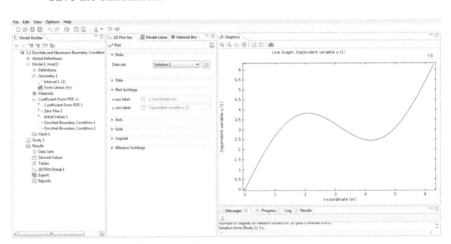

**FIGURE 3.1**    Solution to Equation (3.1) for the Dirichlet boundary condition.

*Step 6*

- Select the *Dirichlet boundary condition* option at *r: 2\*pi* and change it with *Neumann boundary condition* $\dfrac{du}{dx} = 3$ as given in Equation (3.2). At this point, put *r: u + ux − 3*. This step will add Newman condition $\dfrac{du}{dx} = 3$, that is, *ux = 3* (Note: as *u = r, r = u + ux − 3, ux = 0*).

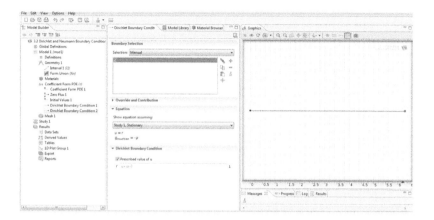

*Step 7*

- Click on the *Mesh* option in *Model Builder.* Select *Normal Mesh* Type. Click on *Build All* option at the top of ribbon. A dialogue box will appear in the *Message* tab as: *"Complete mesh consists of 15 elements."*

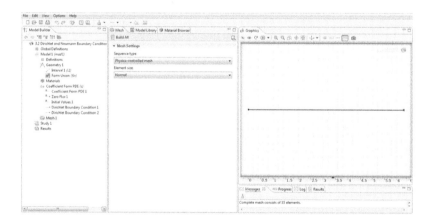

- Now, go to *Study* option in the model pellet tab. Click on the *Compute (=)* button. A graph will appear giving the profile of $u(x)$ versus $x$ as shown in Figure 3.2.
- Save the simulation.

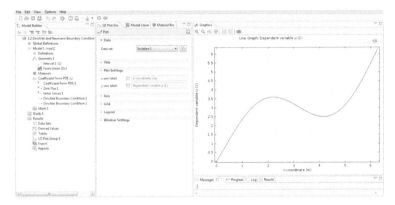

**FIGURE 3.2**    Solution to Equation (3.1) for the Neumann boundary condition.

## 3.3    SIMULATION OF HEAT TRANSFER IN THE RECTANGULAR FIN

### *3.3.1    PROBLEM STATEMENT*

Figure 3.3 shows steady-state heat transfer in the rectangular fin (Ghoshdastidar, 1998). The base temperature of the fin is $T(0) = 0$ and the temperature at the fin tip is $T(1) = 1$.

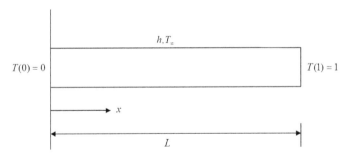

**FIGURE 3.3**    Steady-state heat transfer in rectangular fin.

$h$ = fin heat transfer coefficient
$L$ = fin length

The dimensionless form of the energy equation is

$$\frac{d}{dX}\left(k(T)\frac{dT}{dX}\right) = Q \qquad (3.5)$$

$$T(0) = 0,\ T(1) = 1. \qquad (3.6)$$

The temperature, *T*, is the dimensionless temperature and is a function of dimensionless length, *X*. The thermal conductivity of the fin is temperature dependent $k(T) = 1+T$, $Q=0$.

$Q$ = energy generation rate.

### 3.3.2   *SIMULATION APPROACH (METHOD 1)*

*Step 1*

- Open COMSOL Multiphysics.
- Select *1D Space Dimension* from the list of options. Hit the *next* arrow at the upper right corner.

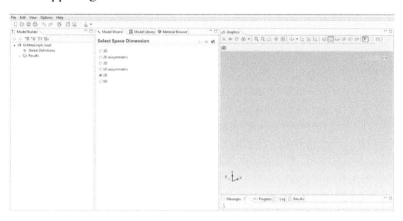

- Select and expand the *Mathematics* folder from the list of options in *Model Wizard*. Further select and expand *PDE Interface* and click on the *Coefficient Form PDE (c)*. Again hit the *next* arrow.

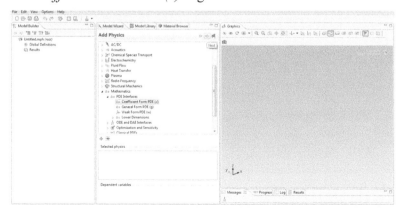

- Then, select *Stationary* from the list of *Study Type* options, and click
  on the *Finish flag* at the upper right corner of the application.

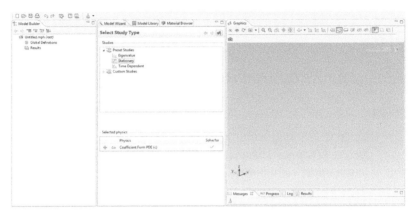

## Step 2

- With the *Model Builder*, right click on the *Geometry* option. From the
  list of options, click on the *Interval* taskbar.
- In the *Interval* environment, select *Intervals*1, *Left endpoint: 0 and
  Right endpoint: 1*. Click on the *Build All* option at the upper part
  of tool bar. A straight line graph will appear on the right side of the
  application window.

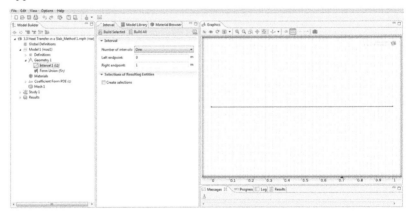

## Step 3

- In the *Model Builder*, select *Coefficient Form PDE* node and change
  the dependent variables to T.

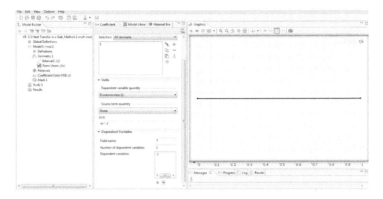

- Now, in the *Model Builder*, select *Coefficient Form PDE 1* option. This will open a tab to enter the coefficients of the characteristics of a model equation.
- In the *Domain Selection* panel, you will see an equation of the form

$$e_a \frac{\partial^2 T}{\partial t^2} + d_a \frac{\partial T}{\partial t} + \nabla \cdot \left(-c\nabla T - \alpha T + \gamma\right) + \beta \cdot \nabla T + aT = f \qquad (3.7)$$

where $\nabla = \left[\dfrac{\partial}{\partial x}\right]$.

In order to solve the governing differential equation, we need to assign the coefficients in above equation a suitable value.

To convert Equation (3.7) to the desired form of Equation (3.5), the value of coefficients in Equation (3.7) to be changed as follows:

$c = -(1+T)$ $\alpha = 0$, $\beta = 0$, $\gamma = 0$
$a = 0 f = 0$
$e_a = 0 d_a = 0$.

This adjustment will reduce Equation (3.7) to Equation (3.5) format.

*Step 4*

- Now, right click on the *Coefficient Form PDE* tab in *Model Builder*. Select the *Dirichlet Boundary Condition* option. Another window will open namely *Dirichlet Boundary Condition 1*. Click on it.
- Select the left point on the horizontal line graph. Click on the "+" sign at the top right corner, this will add *boundary 1* in the *Boundary Selection* tab. At this point, put *r: 0*. This step will add a *Dirichlet boundary condition*: $T(0) = 0$ as given in Equation (3.6) (Note: as $T = r$, $r = 0$, $T = 0$).

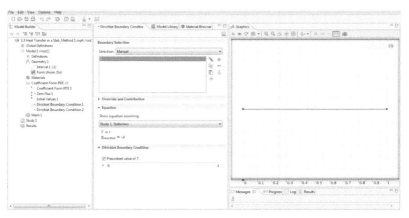

- Again right click on the *Coefficient Form PDE* tab in *Model Builder*. Select the *Dirichlet Boundary Condition* option again. Another window will open namely *Dirichlet Boundary Condition 2*. Click on it.
- Select the right point on the horizontal line graph. Click on the "+" sign at the top right corner, this will add *boundary 2* in the *Boundary Selection* tab. At this point, put *r: 1*. This step will add a *Dirichlet Boundary Condition*: $T(1)=1$ as given in Equation (3.6) (Note: as $T = r$, $r = 1$, $T = 1$).

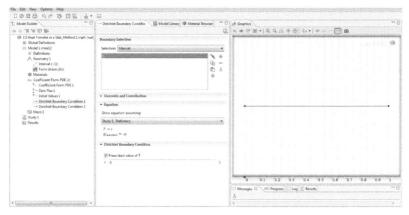

*Step 5*

- Click on the *Mesh* option in *Model Builder.* Select *Normal Mesh* Type. Click on *Build All* option at the top of ribbon. A dialogue box will appear in the *Message* tab as: *"Complete mesh consists of 15 elements."*

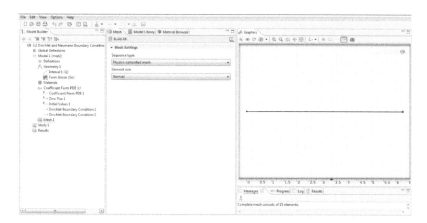

- Now, go to *Study* option in the model pellet tab. Click on the *Compute (=)* button. A graph will appear giving the profile of *T(x)* versus *x* as shown in Figure 3.4.
- Save the simulation.

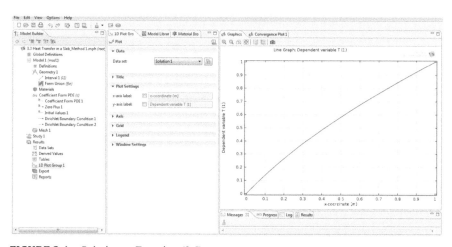

**FIGURE 3.4**  Solution to Equation (3.5).

### 3.3.3   SIMULATION APPROACH (METHOD 2)

*Step 1*

- Open COMSOL Multiphysics.
- Select *1D Space Dimension* from the list of options. Hit the *next* arrow at the upper right corner.

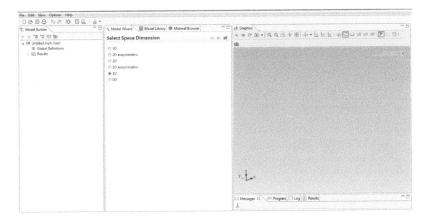

- Select and expand the *Heat Transfer* folder from the list of options in *Model Wizard*. Further select and *Heat Transfer in Solids*. Again hit the *next* arrow.

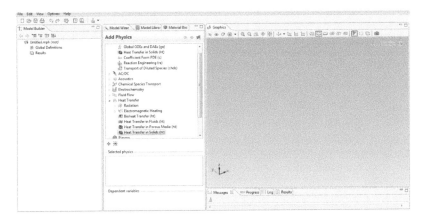

- Then, select *Stationary* from the list of *Study Type* options, and click on the *Finish flag* at the upper right corner of the application.

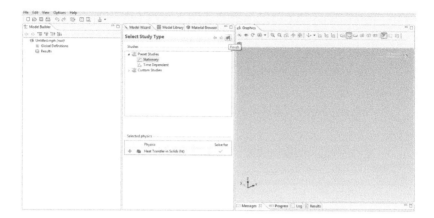

## Step 2

- With the *Model Builder*, right click on the *Geometry* option. From the list of options, click on the *Interval* taskbar.
- In the *Interval* environment, select *Intervals*1, *Left endpoint: 0 and Right endpoint: 1*. Click on the *Build All* option at the upper part of tool bar. A straight line graph will appear on the right side of the application window.

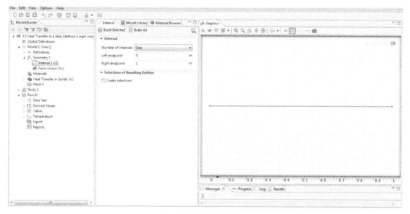

- Now, in the *Model Builder*, select *Heat Transfer in Solids 1* option. This will open a tab to enter the coefficients of the characteristics of a model equation.
- In the *Domain Selection* panel, you will see an equation of the form

$$\tilde{n}C_p u.\nabla T = \nabla \cdot (k\nabla T)) + Q \qquad (3.8)$$

where $\nabla = \left[ \dfrac{\partial}{\partial x} \right]$.

In order to solve the governing differential equation, we need to assign the coefficients in the above equation a suitable value.

To convert Equation (3.8) to the desired form of Equation (3.5), the value of coefficients in Equation (3.8) to be changed as follows:

$k=1+T.$

This adjustment will reduce Equation (3.8) to Equation (3.5) format.

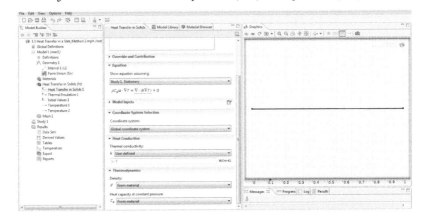

*Step 4*

- In the *Model Builder*, right click on the *Heat Transfer in Solids* tab and select *Temperature* option. Another window will open namely *Temperature 1.* Click on it.
- Select the left point on the horizontal line graph. Click on the "+" sign at the top right corner, this will add *boundary 1* in the *Boundary Selection* tab. At this point, put $T_0$: *0.* This step will add a *Dirichlet boundary condition*: $T(0) = 0$ as given in Equation (3.6).

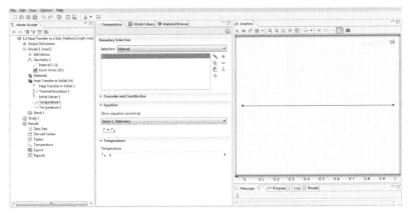

- In the *Model Builder*, again right click on the *Heat Transfer in Solids* tab and select *Temperature* option again. Another window will open namely *Temperature 2*. Click on it.
- Select the right point on the horizontal line graph. Click on the "+" sign at the top right corner, this will add *boundary 2* in the *Boundary Selection* tab. At this point, put $T_0$: *1*. This step will add a *Dirichlet boundary condition*: $T(1)=1$ as given in Equation (3.6).

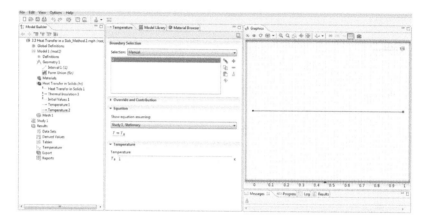

*Step 5*

- Click on the *Mesh* option in *Model Builder*. Select *Normal Mesh* Type. Click on *Build All* option at the top of ribbon. A dialogue box will appear in the *Message* tab as: "*Complete mesh consists of 15 elements.*"

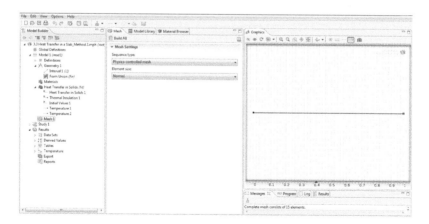

- Now, go to *Study* option in the model pellet tab. Click on the *Compute (=)* button. A graph will appear giving the profile of *T(x)* versus *x* as shown in Figure 3.5.
- Save the simulation.

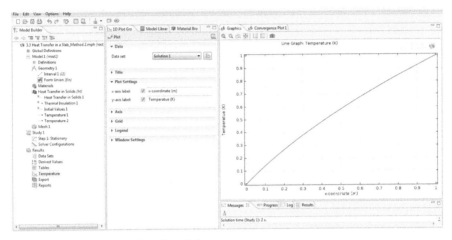

**FIGURE 3.5**    Solution to Equation (3.5).

*Step 6*

- To plot the heat flux, select the *Line Graph* in the *Temperature* tab. Another *Plot* window will open. Click on the "+" sign in the *y*-axis data and select *Conductive Heat Flux, x* component.

- Click on *Plot* icon in the *Line Graph*. A graph will appear giving the profile of *Conductive Heat Flux* versus *x* as shown in Figure 3.6.

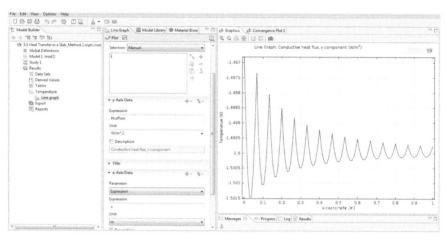

**FIGURE 3.6** Plot of heat flux in Equation (3.5).

## Step 7

- To plot the thermal conductivity, select the *Line Graph* in the *Temperature* tab. Another *Plot* window will open. Click on the "+" sign in the *y*-axis data and select *Thermal Conductivity, xx* component.

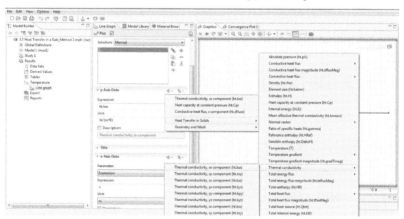

- Click on *Plot* icon in the *Line Graph*. A graph will appear giving the profile of *Thermal Conductivity* versus *x* as shown in Figure 3.7.

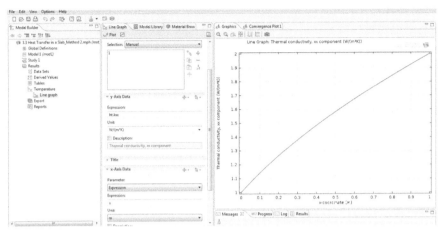

**FIGURE 3.7**    Thermal conductivity in Equation (3.5).

## 3.4    SIMULATION OF REACTION–DIFFUSION

### *3.4.1    PROBLEM STATEMENT*

The illustration on reaction–diffusion considering spherical domain (Finlayson, 2006) is given by following differential equation:

$$\frac{1}{r^2}\frac{d}{dr}\left(r^2\frac{dc}{dr}\right) = \frac{\alpha'c}{1+Kc} \tag{3.9}$$

$$\text{or} \ -\frac{d^2c}{dr^2} = \frac{2}{r}\frac{dc}{dr} - \frac{\alpha'c}{1+Kc}. \tag{3.10}$$

The boundary conditions are

$$\frac{dc}{dr}(0) = 0, c(1) = 1 \tag{3.11}$$

where $\alpha = 5$ and $K = 2$.

### *3.4.2    SIMULATION APPROACH (METHOD 1)*

*Step 1*

- Open COMSOL Multiphysics.
- Select *1D Space Dimension* from the list of options. Hit the *next* arrow at the upper right corner.

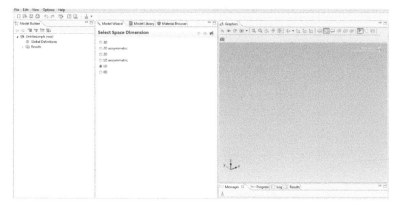

- Select and expand the *Mathematics* folder from the list of options in *Model Wizard*. Further select and expand *PDE Interface* and click on the *Coefficient Form PDE (c)*. Again hit the *next* arrow.

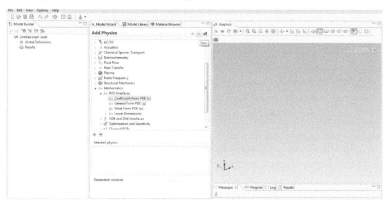

- Then, select *Stationary* from the list of *Study Type* options, and click on the *Finish flag* at the upper right corner of the application.

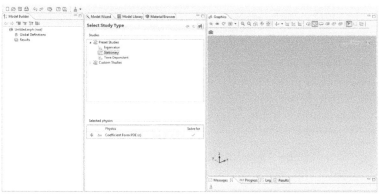

*Step 2*

- With the *Model Builder*, right click on the *Geometry* option. From the list of options, click on the *Interval* taskbar.
- In the *Interval* environment, select *Intervals* 1, *Left endpoint: 0 and Right endpoint:1*. Click on the *Build All* option at the upper part of tool bar. A straight line graph will appear on the right side of the application window.

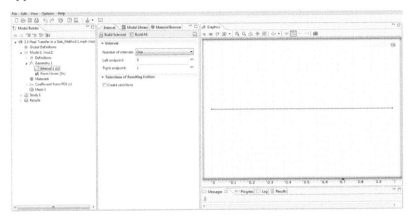

*Step 3*

- In the *Model Builder*, select *Coefficient Form PDE* node and change the Dependent variables to "*cn*".

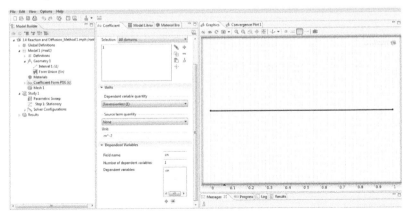

- Now, in the *Model Builder*, select *Coefficient Form PDE 1* option. This will open a tab to enter the coefficients of the characteristics of a model equation.

- In the *Domain Selection* panel, you will see an equation of the form

$$e_a \frac{\partial^2 cn}{\partial t^2} + d_a \frac{\partial cn}{\partial t} + \nabla \cdot \left( -c\nabla cn - \alpha cn + \gamma \right) + \beta \cdot \nabla cn + acn = f \quad (3.12)$$

where $\nabla = \left[ \dfrac{\partial}{\partial x} \right]$.

In order to solve the governing differential equation, we need to assign the coefficients in above equation a suitable value.

To convert Equation (3.12) to the desired form of Equation (3.10), the value of coefficients in Equation (3.12) to be changed as follows:
$c = 1$ $\alpha = 0$, $\beta = -(2/x)$, $\gamma = 0$
$a = 0$ $f = -(alpha*cn)/(1+K*cn)$
$e_a = 0$ $d_a = 0$.

This adjustment will reduce Equation (3.12) to Equation (3.10) format.

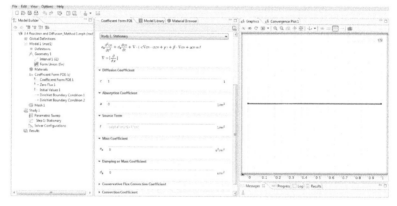

- In the *Model Builder*, right click on the *Definitions* tab and select *Variables*. Another window will open namely *Variables 1*. Click on it and define *K: 2* and *alpha: 5*.

*Step 4*

- Now, right click on the *Coefficient Form PDE* tab in *Model Builder.* Select the *Dirichlet Boundary Condition* option. Another window will open namely *Dirichlet Boundary Condition 1*. Click on it.
- Select the left point on the horizontal line graph. Click on the "+" sign at the top right corner, this will add *boundary 1* in the *Boundary Selection* tab. At this point, put *r: cn+cnx*. This step will add the *Neumann Boundary Condition* $\dfrac{dc}{dr}(0)=0$, that is, $\dfrac{dcn}{dx}(0)=0$, that is, *cnx=0* as given in Equation (3.11) (Note: as *cn = r, r = cn + cnx, cnx = 0*).

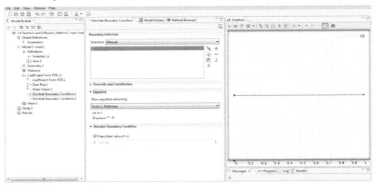

- Again right click on the *Coefficient Form PDE* tab in *Model Builder.* Select the *Dirichlet Boundary Condition* option again. Another window will open namely *Dirichlet Boundary Condition 2*. Click on it.
- Select the right point on the horizontal line graph. Click on the "+" sign at the top right corner, this will add *boundary 2* in the *Boundary Selection* tab. At this point, put *r: 1*. This step will add a *Dirichlet boundary condition: c(1)=1*, that is, *cn(1)=1* as given in Equation (3.11) (Note: as *cn = r, r = 1, cn = 1*).

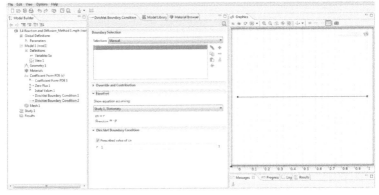

*Step 5*

- Click on the *Mesh* option in *Model Builder.* Select *Normal Mesh* Type. Click on *Build All* option at the top of ribbon. A dialogue box will appear in the *Message* tab as: *"Complete mesh consists of 15 elements."*

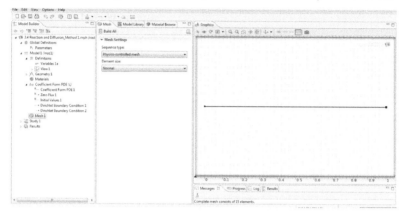

- Now, go to *Study* option in the model pellet tab. Click on the *Compute (=)* button. A graph will appear giving the profile of *Concentration* versus *r*.
- Save the simulation.

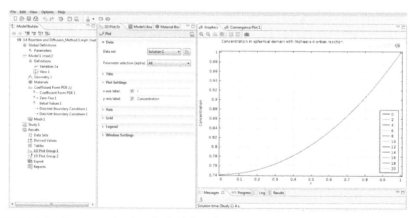

**FIGURE 3.8**  Concentration in spherical domain [solution to Equation (3.10)].

### 3.4.2.1  *SOLUTION FOR SEVERAL VALUES OF A*

*Step 1*

- In the *Model Builder*, right click on the *Global Definitions* tab and select *Parameters*. Another window will open namely *Parameters*.

Click on it and define *alpha: 5*. Then, remove alpha from *Variables* under *Definitions*.

*Step 2*

• Now, right click on the *Study 1* tab in *Model Builder*. Select the *Parametric Sweep* option. Another window will open namely *Study Settings*. Click on "+" sign to add "alpha" in the *Parameter names* and set the range(0,2,20).

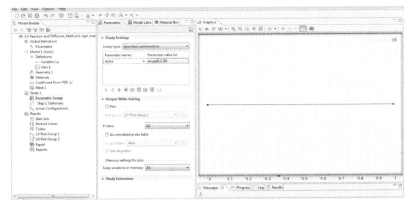

• Now go to *Study* option in the model pellet tab. Click on the *Compute (=)* button. A graph will appear giving the profile of *Concentration* versus *r*.
• Save the simulation.

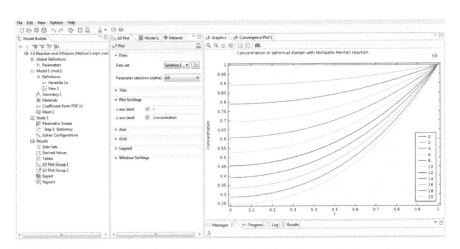

**FIGURE 3.9**    Concentration in spherical domain [solution for several values of α].

### 3.4.3   SIMULATION APPROACH (METHOD 2)

*Step 1*

- Open COMSOL Multiphysics.
- Select *1D Space Dimension* from the list of options. Hit the *next* arrow at the upper right corner.

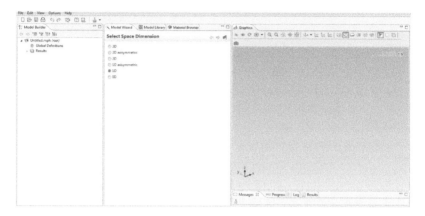

- Expand the *Chemical Species Transport* folder from the list of options in *Model Wizard* and select *Transport of Diluted Species*. Hit the *next* arrow again.

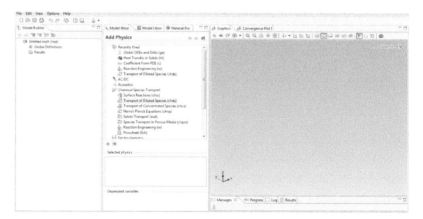

- Then, select *Stationary* from the list of *Study Type* options, and click on the *Finish flag* at the upper right corner of the application.

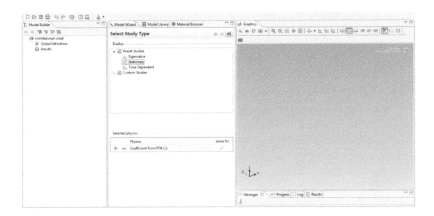

### Step 2

- With the *Model Builder*, right click on the *Geometry* option. From the list of options, click on the *Interval* taskbar.
- In the *Interval* environment, select *Intervals*1, *Left endpoint: 0 and Right endpoint: 1*. Click on the *Build All* option at the upper part of tool bar. A straight line graph will appear on the right side.

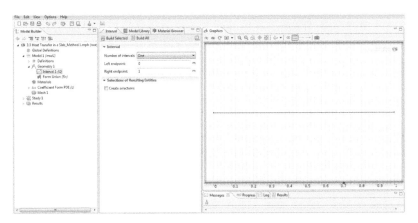

### Step 3

- In the *Model Builder*, select the *Transport of Diluted Species* tab and expand the Dependent variables tab. Change the species concentration to "*c*." Choose the *Discretization tab* and change the *Concentration* from *Linear* to *Quadratic*.

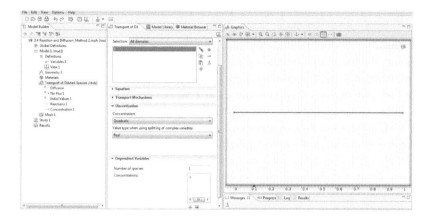

- Now, in the *Model Builder*, select *Diffusion* option available in the *Transport of Diluted Species*. This will open a tab to enter the coefficients of the characteristics of a model equation.
- In the *Domain Selection* panel, you will see an equation of the form

$$\nabla \cdot \left( -D_i \nabla c_i \right) = R_i \qquad (3.13)$$

where $N_i = -D_i \nabla c_i$.

In order to solve the governing differential equation, we need to assign the coefficients in above equation a suitable value.

To convert Equation (3.13) to the desired form of Equation (3.10), the value of coefficients in Equation (3.13) to be changed as follows: Change the value of the *Diffusion coefficient* $D_c$ to 1.

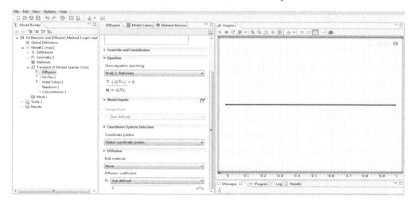

- Select *Reactions1* option available in the *Transport of Diluted Species* tab and expand the Reactions tab. Change the $R_c$ to *(2/x)\*cx – (alpha\*c)/*

*(1+K\*c)* [Note: $\dfrac{2}{r}\dfrac{dc}{dr} = \dfrac{2}{x}\dfrac{dc}{dx} = (2/x)*cx$]. Select the horizontal line graph. Click on the "+" sign at the top right corner, this will add a horizontal line in the *Domain Selection* tab.

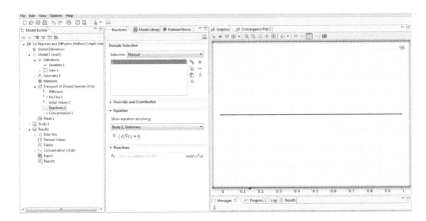

- In the *Model Builder*, right click on the *Definitions* tab and select *Variables*. Another window will open namely *Variables 1*. Click on it and define *K: 2* and *alpha: 5*.

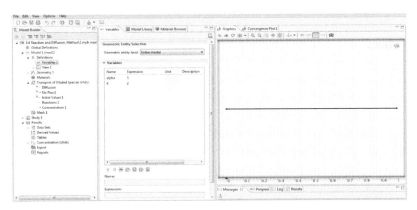

*Step 4*

- In the *Model Builder*, right click on the *Transport of Diluted Species* tab and select *Concentration* option. Another window will open namely *Concentration 1*. Click on it.

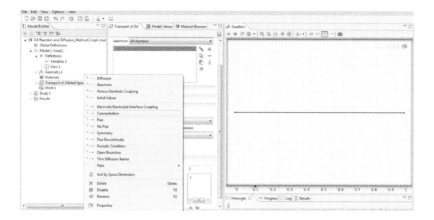

- Select the right point on the horizontal line graph. Click on the "+" sign at the top right corner; this will add *boundary 2* in the *Boundary Selection* tab. At this point, put $c_{0,c}$: *1*. This step will add a *Dirichlet boundary condition*: *c(1)=1*.

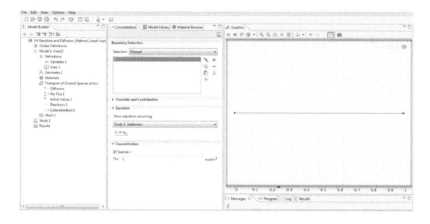

- Now, select *No Flux 1* option available in the *Transport of Diluted Species*. Then, select the left point on the horizontal line graph. The equation at this point is: $-n.N_i = 0$. This is equivalent to $\dfrac{dc}{dr}(0) = 0$ as given in the equation.

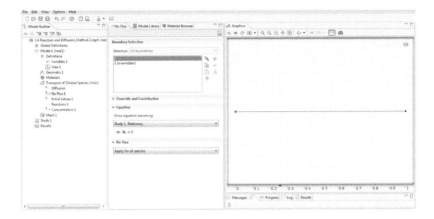

## Step 5

- Click on the *Mesh* option in *Model Builder.* Select *Normal Mesh* Type. Click on *Build All* option at the top of ribbon. A dialogue box will appear in the *Message* tab as: *"Complete mesh consists of 15 elements"*.

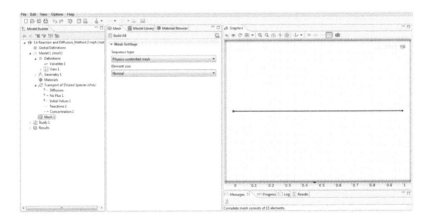

- Now, go to *Study* option in the model pellet tab. Click on the *Compute (=)* button. A graph will appear giving the profile of *Concentration versus r*.
- Save the simulation.

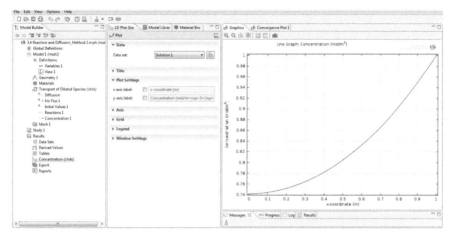

**FIGURE 3.10**  Solution to Equation (3.10).

## 3.5  SIMULATION OF NEWTONIAN FLUID IN A PIPE

### 3.5.1  *PROBLEM STATEMENT*

Figure 3.11 illustrates the pipe of radius $R$ and length $L$ over which pressure drop $\Delta p$ occurs. The Newtonian fluid having viscosity $\mu$ is flowing through the pipe with velocity $v$. It is governed by the following differential equation (Bird et al., 2002):

$$\frac{d^2v}{dr^2} = -\frac{1}{\mu}\frac{\Delta p}{L} - \frac{1}{r}\frac{dv}{dr} \tag{3.14}$$

$$\frac{dv}{dr}(0) = 0, v(R) = 0 \tag{3.15}$$

where

$\Delta p = 28 \times 10^4$ Pa

$\mu = 492 \times 10^{-3}$ Pa s

$L = 488 \times 10^{-2}$ m

$R = 0.0025$ m.

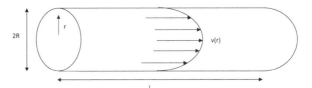

**FIGURE 3.11**    Newtonian fluid in a pipe.

## 3.5.2   *SIMULATION APPROACH (METHOD 1)*

*Step 1*

- Open COMSOL Multiphysics.
- Select *1D* axisymmetric *Space Dimension* from the list of options. Hit the *next* arrow at the upper right corner.

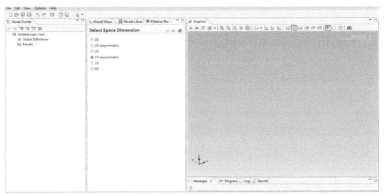

- Select and expand the *Mathematics* folder from the list of options in *Model Wizard*. Further select and expand *PDE Interface* and click on the *Coefficient Form PDE (c)*. Again hit the *next* arrow.

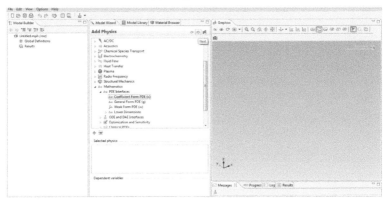

- Then, select *Stationary* from the list of *Study Type* options, and click on the *Finish flag* at the upper right corner of the application.

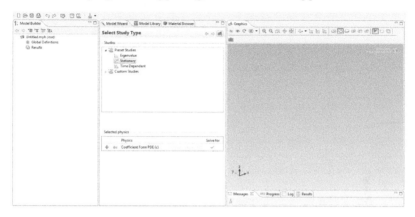

*Step 2*

- With the *Model Builder*, right click on the *Geometry* option. From the list of options, click on the *Interval* taskbar.
- In the *Interval* environment, select *Intervals 1, Left endpoint: 0 and Right endpoint: 0.0025*. Click on the *Build All* option at the upper part of tool bar. A straight line graph will appear on the right side of the application window.

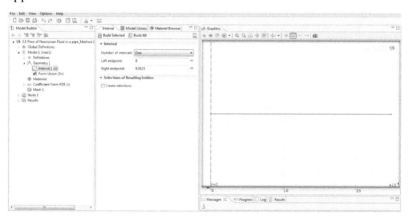

*Step 3*

- In the *Model Builder*, select *Coefficient Form PDE* node and change the Dependent variables to "*v*".

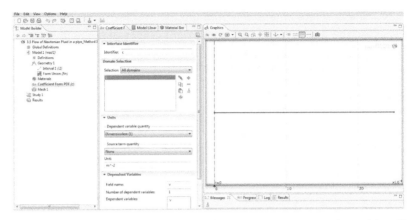

- Now in the *Model Builder*, select *Coefficient Form PDE 1* option. This will open a tab to enter the coefficients of the characteristics of a model equation.
- In the *Domain Selection* panel, you will see an equation of the form

$$e_a \frac{\partial^2 v}{\partial t^2} + d_a \frac{\partial v}{\partial t} + \nabla \cdot \left( -c\nabla v - \alpha v + \gamma \right) + \beta . \nabla v + av = f \qquad (3.16)$$

where $\nabla = \left[ \dfrac{\partial}{\partial r} \right]$.

In order to solve the governing differential equation, we need to assign the coefficients in above equation a suitable value.

To convert Equation (3.16) to the desired form of Equation (3.14), the value of coefficients in Equation (3.16) to be changed as follows:

$c = -1 \ \alpha = 0, \ \beta = 0, \ \gamma = 0$

$a = 0 \ f = -dpL - vr/r$

$e_a = 0 \ d_a = 0.$

This adjustment will reduce Equation (3.16) to Equation (3.14) format.

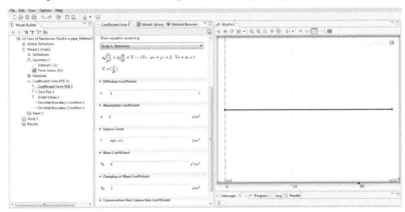

- In the *Model Builder*, right click on the *Global Definitions* tab and select *Variables*. Another window will open namely *Variables 1*. Click on it and define *dpL: dp/(mu\*L)*, *dp: 2.8e5*, *mu: 0.492*, and *L: 4.88*.

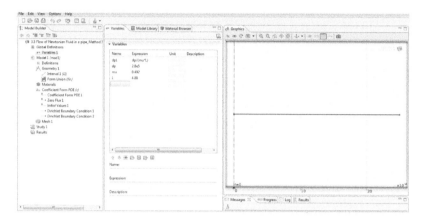

*Step 4*

- Now, right click on the *Coefficient Form PDE* tab in *Model Builder*. Select the *Dirichlet Boundary Condition* option. Another window will open namely *Dirichlet Boundary Condition 1*. Click on it.
- Select the left point on the horizontal line graph. Click on the "+" sign at the top right corner, this will add *boundary 1* in the *Boundary Selection* tab. At this point, put *r: v+vr*. This step will add the *Neumann Boundary Condition* $\frac{dv}{dr}(0) = 0$, that is, *vr = 0* as given in Equation (3.15) (Note: as *v = r, r = v + vr, vr = 0*).

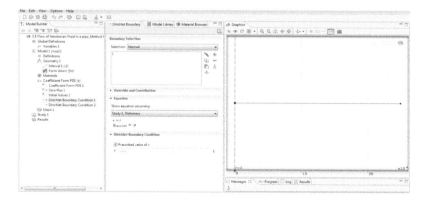

- Again right click on the *Coefficient Form PDE* tab in *Model Builder.* Select the *Dirichlet Boundary Condition* option again. Another window will open namely *Dirichlet Boundary Condition 2.* Click on it.
- Select the right point on the horizontal line graph. Click on the "+" sign at the top right corner, this will add *boundary 2* in the *Boundary Selection* tab. At this point, put *r: 0.* This step will add a *Dirichlet boundary condition: v(R) = 0* as given in Equation (3.15) (Note: as *v = r, r = 0, v = 0*).

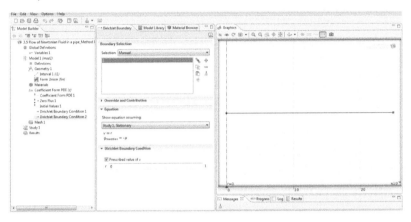

*Step 5*

- Click on the *Mesh* option in *Model Builder.* Select *Normal Mesh* Type. Click on *Build All* option at the top of ribbon. A dialogue box will appear in the *Message* tab as: *"Complete mesh consists of 15 elements"*.

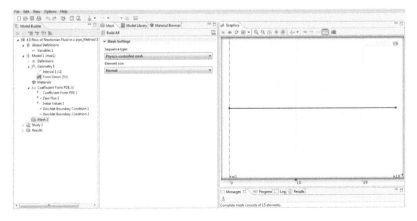

- Now, go to *Study* option in the model pellet tab. Click on the *Compute (=)* button. A graph will appear giving the profile of *Velocity versus Radius* as shown in Figure 3.12.
- Save the simulation.

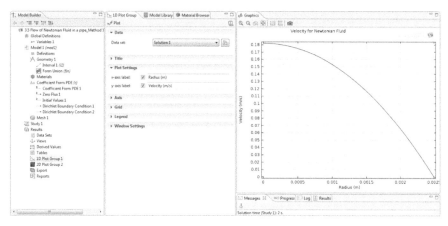

**FIGURE 3.12**   Solution to Equation (3.14).

*Step 6*

- In the *Model Builder*, right click on the *Results* tab and select the *2D Plot Group* option. Another window will open namely *2D Plot Group 2*. Right click on it and select *Surface*. Another window will open namely *Surface 1*. Then select *Height Expression* by right clicking on it. A 3D plot appears.

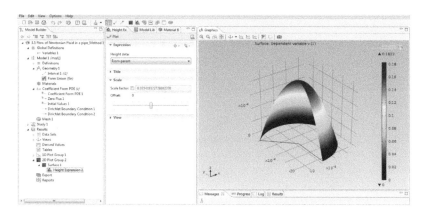

### 3.5.3   SIMULATION APPROACH (METHOD 2)

*Step 1*

- Open COMSOL Multiphysics.
- Select *1D* axisymmetric *Space Dimension* from the list of options. Hit the *next* arrow at the upper right corner.

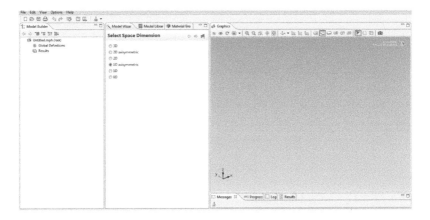

- Select and expand the *Mathematics* folder from the list of options in *Model Wizard* and click on the *Classical PDEs* and select the *Poisson's equation*. Again hit the *next* arrow.

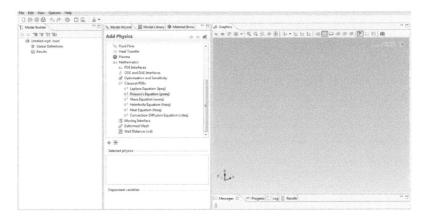

- Then, select *Stationary* from the list of *Study Type* options, and click on the *Finish flag* at the upper right corner of the application.

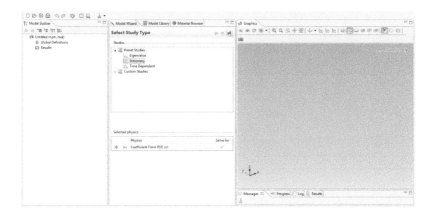

## Step 2

- With the *Model Builder*, right click on the *Geometry* option. From the list of options, click on the *Interval* taskbar.
- In the *Interval* environment, select *Intervals*1, *Left endpoint: 0 and Right endpoint: 0.0025*. Click on the *Build All* option at the upper part of tool bar. A straight line graph will appear on the right side of the application window.

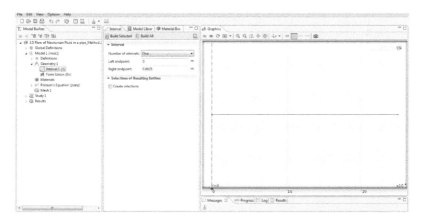

## Step 3

- In the *Model Builder*, select *Poisson's equation* node and change the Dependent variables to "*v*."

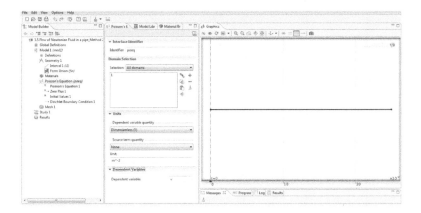

- Now, in the *Model Builder*, select *Poisson's equation 1* option available in *Poisson's equation*. This will open a tab to enter the coefficients of the characteristics of a model equation.
- In the *Domain Selection* panel, you will see an equation of the form

$$\nabla.\left(-c\nabla v\right) = f \tag{3.17}$$

where $\nabla = [\dfrac{\partial}{\partial r}]$

In order to solve the governing differential equation, we need to assign the coefficients in above equation a suitable value.

To convert Equation (3.17) to the desired form of Equation (3.14), the value of coefficients in Equation (3.17) to be changed as follows:

$c = -1\ \alpha = 0,\ \beta = 0,\ \gamma = 0$

$a = 0\ f = -dpL - vr/r$

$e_a = 0\ d_a = 0.$

This adjustment will reduce Equation (3.17) to Equation (3.14) format.

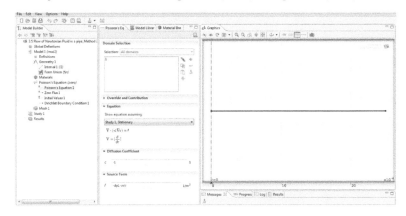

- In the *Model Builder*, right click on the *Global Definitions* tab and select *Variables*. Another window will open namely *Variables 1*. Click on it and define *dpL: dp/(mu\*L)*, *dp: 2.8e5*, *mu: 0.492*, and *L: 4.88*.

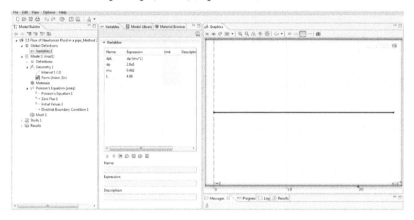

*Step 4*

- Now, click again on the *Poisson's Equation* tab in *Model Builder*. Select the *Dirichlet Boundary Condition* option. Another window will open namely *Dirichlet Boundary Condition 1*. Click on it.
- Select the right point on the horizontal line graph. Click on the "+" sign at the top right corner; this will add *boundary 2* in the *Boundary Selection* tab. At this point, put *r: 0*. This step will add a *Dirichlet boundary condition*: $v(R) = 0$ as given in Equation (3.15) (Note: as $v = r$, $r = 0$, $v = 0$).

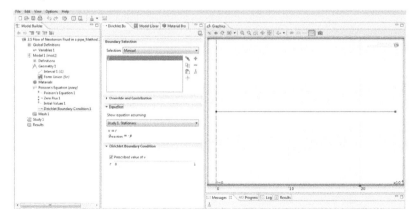

- Now, select *Zero Flux 1* option available in *Poisson's equation*. The equation at this point is: $-n.(-c\nabla v) = 0$. This is equivalent to $\dfrac{dv}{dr}(0) = 0$ as given in Equation (3.15).

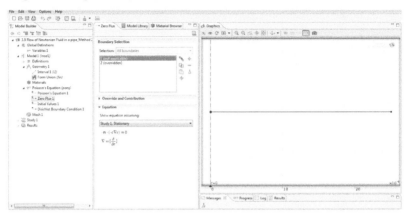

*Step 5*

- Click on the *Mesh* option in *Model Builder*. Select *Normal Mesh* Type. Click on *Build All* option at the top of ribbon. A dialogue box will appear in the *Message* tab as: *"Complete mesh consists of 15 elements"*.

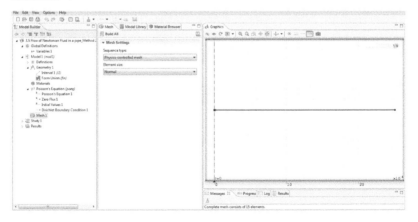

- Now, go to *Study* option in the model pellet tab. Click on the *Compute (=)* button. A graph will appear giving the profile of *Velocity versus Radius* as shown in Figure 3.13.
- Save the simulation.

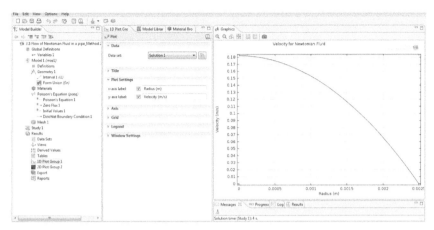

**FIGURE 3.13**    Solution to Equation (3.14).

*Step 6*

- Now, right click on the *Results* tab in *Model Builder.* Select the *2D Plot Group* option. Another window will open namely *2D Plot Group 2.* Right click on *2D Plot Group 2* and *Surface.* Another window will open namely *Surface 1.* Right click on it and choose *Height Expression.* A 3D plot appears.

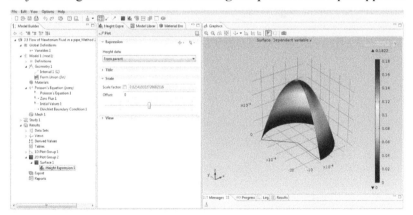

## 3.6    SIMULATION OF NON-NEWTONIAN FLUID IN A PIPE

### 3.6.1    *PROBLEM STATEMENT*

The non-Newtonian fluid having viscosity $\eta$ (Bird et al., 1987) is flowing through the pipe with velocity $v$. It is governed by the following differential equation:

$$\frac{Ä}{r}\frac{d}{dr}\left(r\eta\frac{dv}{dr}\right) = -\frac{\Delta P}{L} \; or \; \frac{d}{dr}\left(\eta\frac{dv}{dr}\right)+\frac{\eta}{r}\frac{dv}{dr} = -\frac{P}{L} \tag{3.18}$$

$$\frac{dv}{dr}(0) = 0, v(R) = 0 \tag{3.19}$$

$$\eta = \frac{\eta_0}{\left[1+\left(\lambda\frac{dv}{dr}\right)^2\right]^{(1-n)/2}} \tag{3.20}$$

where

$\eta_0 = 0.492$

$\lambda = 0.1$

$n = 0.8$

$\Delta p = 2.8 \times 10^5$ Pa

$L = 488$ m

$R = 0.0025$ m.

### 3.6.2   SIMULATION APPROACH

*Step 1*

- Open COMSOL Multiphysics.
- Select *1D* axisymmetric *Space Dimension* from the list of options. Hit the *next* arrow at the upper right corner.

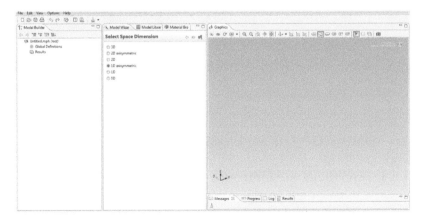

- Select and expand the *Mathematics* folder from the list of options in *Model Wizard*. Further select and expand *PDE Interface* and click on the *Coefficient Form PDE (c)*. Again hit the *next* arrow.

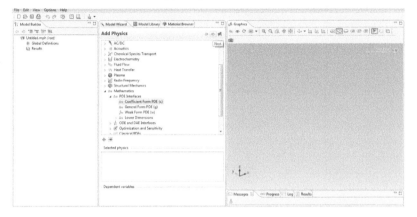

- Then, select *Stationary* from the list of *Study Type* options, and click on the *Finish flag* at the upper right corner of the application.

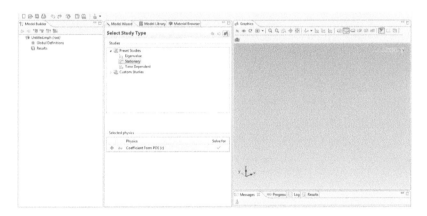

*Step 2*

- With the *Model Builder*, right click on the *Geometry* option. From the list of options, click on the *Interval* taskbar.
- In the *Interval* environment, select *Intervals1, Left endpoint: 0 and Right endpoint: 0.0025*. Click on the *Build All* option at the upper part of tool bar. A straight line graph will appear on the right side of the application window.

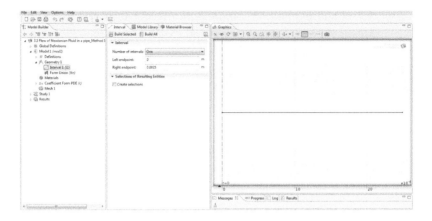

## Step 3

- In the *Model Builder*, select *Discretization tab* available under *Show*. Then, go to *Coefficient Form PDE* node and change the Dependent variables to "*v*." Choose the *Discretization tab* and change the *Shape function type* to *Lagrange* and *Element order* to *Quadratic*.

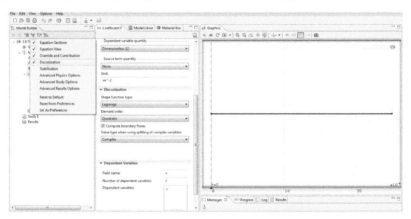

- Now, in the *Model Builder*, select *Coefficient Form PDE 1* option. This will open a tab to enter the coefficients of the characteristics of a model equation.

In the *Domain Selection* panel, you will see an equation of the form

$$e_a \frac{\partial^2 v}{\partial t^2} + d_a \frac{\partial v}{\partial t} + \nabla \cdot \left( -c\nabla v - \alpha v + \gamma \right) + \beta \cdot \nabla v + av = f \qquad (3.21)$$

where $\nabla = \left[ \dfrac{\partial}{\partial r} \right]$.

In order to solve the governing differential equation, we need to assign the coefficients in above equation a suitable value.

To convert Equation (3.21) to the desired form of Equation (3.18), the value of coefficients in Equation (3.18) to be changed as follows:

$c = -eta$ $\alpha = 0,$ $\beta = 0,$ $\gamma = 0$

$a = 0$ $f = -dpL - (eta/r)*vr$

$e_a = 0$ $d_a = 0.$

This adjustment will reduce Equation (3.21) to Equation (3.18) format.

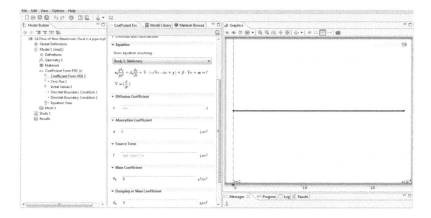

- In the *Model Builder*, right click on the *Definitions* tab and select *Variables*. Another window will open namely *Variables 1*. Click on it and define *dpL: dp/L, dp: 2.8e5, mu: 0.492, eta: (0.492)/$Q^P$, L: 4.88, lambda: 0.1, n: 0.8, Q: (1+(lambda\*vr)$^2$)* and *P: (1 – n)/2.*

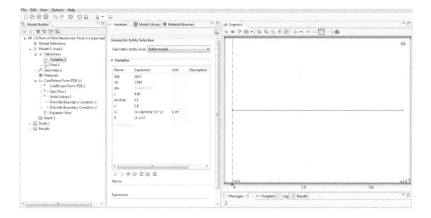

*Step 4*

- Now, right click on the *Coefficient Form PDE* tab in *Model Builder.* Select the *Dirichlet Boundary Condition* option. Another window will open namely *Dirichlet Boundary Condition 1.* Click on it.
- Select the left point on the horizontal line graph. Click on the "+" sign at the top right corner, this will add *boundary 1* in the *Boundary Selection* tab. At this point, put *r: v+vr.* This step will add the *Neumann Boundary Condition* $\frac{dv}{dr}(0) = 0,$ that is, *vr=0* as given in Equation (3.19) (Note: as $v = r,\ r = v + vr,\ vr = 0$).

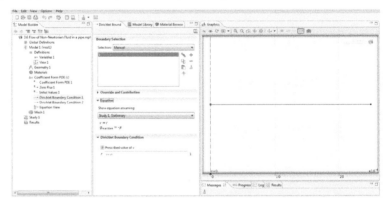

- Again right click on the *Coefficient Form PDE* tab in *Model Builder.* Select the *Dirichlet Boundary Condition* option again. Another window will open namely *Dirichlet Boundary Condition 2.* Click on it.
- Select the right point on the horizontal line graph. Click on the "+" sign at the top right corner, this will add *boundary 2* in the *Boundary Selection* tab. At this point, put *r: 0.* This step will add a *Dirichlet boundary condition:* $v(R) = 0$ as given in Equation (3.19) (Note: as $v = r,\ r = 0,\ v = 0$).

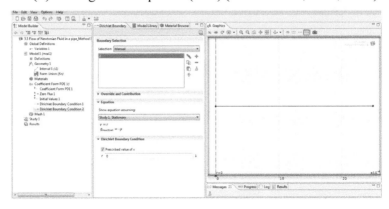

*Step 5*

- Click on the *Mesh* option in *Model Builder.* Select *Normal Mesh*
  Type. Click on *Build All* option at the top of ribbon. A dialogue box
  will appear in the *Message* tab as: *"Complete mesh consists of 15
  elements".*

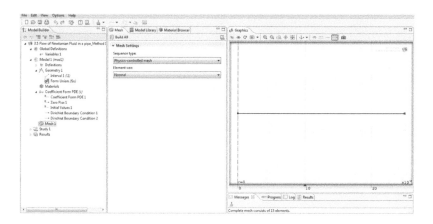

- Now, go to *Study* option in the model pellet tab. Click on the *Compute*
  *(=)* button. A graph will appear giving the profile of *Velocity versus*
  *Radius* as shown in Figure 3.14.
- Save the simulation.

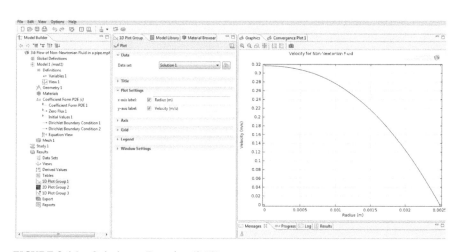

**FIGURE 3.14** Solution to Equation (3.18).

*Step 6*

- In the *Model Builder*, right click on the *Results* tab and select the *2D Plot Group* option. Another window will open namely *2D Plot Group 2*. Right click on it and select *Surface*. Another window will open namely *Surface 1*. Then, select *Height Expression* by right clicking on it. A 3D plot appears.

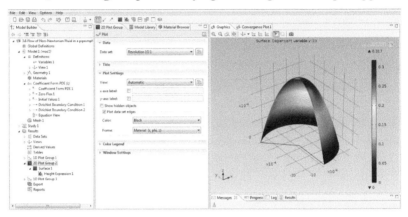

*Step 7*

- To plot the viscosity as a function of radius, select the *Line Graph 1* in the *1D Plot Group 1* tab. Another *Plot* window will open. Change the *Expression* in the *y-Axis Data* from *v* to eta. Now, go to *Study* option in the model pellet tab. Click on the *Compute (=)* button. A graph will appear giving the profile of *Viscosity versus Radius* as shown in Figure 3.15.
- Save the simulation.

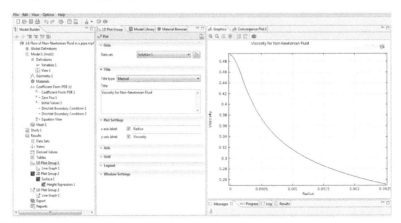

**FIGURE 3.15**   (a) Viscosity [see Equation (3.18)] using quadratic elements for velocity. (b) Viscosity, [see Equation (3.18)] using linear elements for velocity.

- Now, go to the *Discretization tab* available under *Coefficient Form PDE* and change the *Element order* to *Linear* Click on the *Compute (=)* button. A graph will appear giving the profile of *Viscosity versus Radius* as shown in Figure 3.15b.
- Save the simulation.

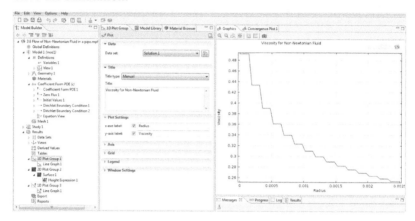

## 3.7   SIMULATION OF TRANSIENT HEAT TRANSFER

### 3.7.1   PROBLEM STATEMENT

Consider a hot infinite plate of finite thickness $L$ as shown in Figure 3.16. The governing differential equation is

$$\frac{\partial \theta}{\partial \tau} = \alpha \frac{\partial^2 \theta}{\partial X^2}, \; \alpha = 2$$

where $\theta = \dfrac{T - T_\infty}{T_i - T_\infty}, \; X = \dfrac{x}{L}, \tau = \dfrac{\alpha t}{L^2}$ \hfill (3.22)

$\theta$, $X$, and $\tau$ are dimensionless temperature, dimensionless distance, and dimensionless time, respectively.

$$\alpha = \frac{k}{\rho C_p}, \text{ thermal diffusivity.} \hfill (3.23)$$

Boundary condition:

$$\frac{\partial \theta}{\partial X}(X = 0) = 0, \theta = 0 \, at \, X = 1 \hfill (3.24)$$

Initial condition:

$$\theta(X, 0) = 1. \tag{3.25}$$

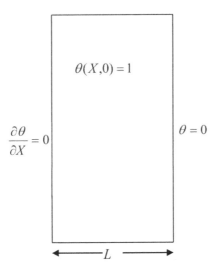

**FIGURE 3.16**   Computational domain of the 1D transient conduction problem.

### 3.7.2   *SIMULATION APPROACH (METHOD 1)*

*Step 1*

- Open COMSOL Multiphysics.
- Select *1D Space Dimension* from the list of options. Hit the *next* arrow at the upper right corner.

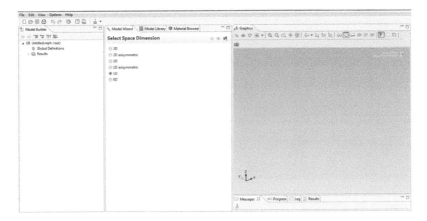

- Select and expand the *Mathematics* folder from the list of options in *Model Wizard*. Further select and expand *PDE Interface* and click on the *Coefficient Form PDE (c)*. Again hit the *next* arrow.

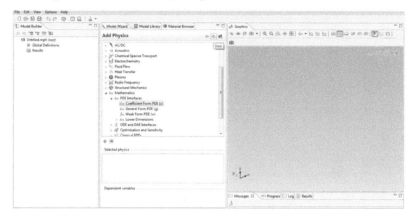

- Then, select *Time Dependent* from the list of *Study Type* options, and click on the *Finish flag* at the upper right corner of the application.

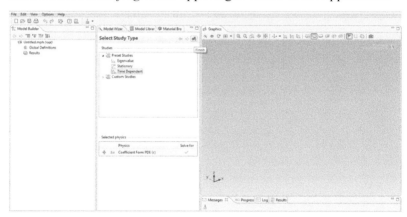

*Step 2*

- With the *Model Builder*, right click on the *Geometry* option. From the list of options, click on the *Interval* taskbar.
- In the *Interval* environment, select *Intervals 1, Left endpoint: 0* and *Right endpoint: 1*. Click on the *Build All* option at the upper part of tool bar. A straight line graph will appear on the right side of the application window.

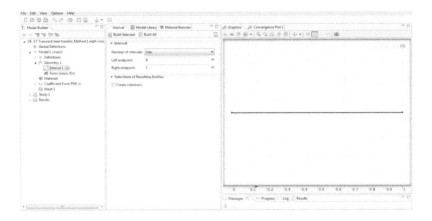

## Step 3

- In the *Model Builder*, select *Coefficient Form PDE* node and change the Dependent variables to T.

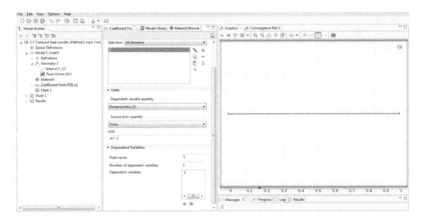

- Now, in the *Model Builder*, select *Coefficient Form PDE 1* option. This will open a tab to enter the coefficients of the characteristics of a model equation.
- In the *Domain Selection* panel, you will see an equation of the form

$$e_a \frac{\partial^2 T}{\partial t^2} + d_a \frac{\partial T}{\partial t} + \nabla.\left(-c\nabla\mathrm{T} - \alpha T + \gamma\right) + \beta.\nabla\mathrm{T} + aT = f \qquad (3.26)$$

where $\nabla = \left[\dfrac{\partial}{\partial x}\right]$

In order to solve the governing differential equation, we need to assign the coefficients in above equation a suitable value.

To convert Equation (3.26) to the desired form of Equation (3.22), the value of coefficients in Equation (3.26) to be changed as follows:

$c = 2\ \alpha = 0,\ \beta = 0,\ \gamma = 0$

$a = 0\ f = 0$

$e_a = 0\ d_a = 1.$

This adjustment will reduce Equation (3.26) to Equation (3.22) format.

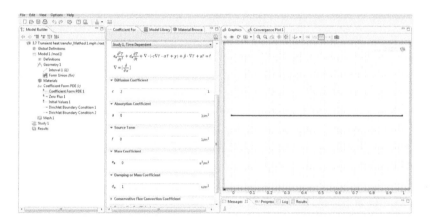

## Step 4

- Now, right click on the *Coefficient Form PDE* tab in *Model Builder*. Select the *Dirichlet Boundary Condition* option. Another window will open namely *Dirichlet Boundary Condition 1*. Click on it.
- Select the left point on the horizontal line graph. Click on the "+" sign at the top right corner, this will add *boundary 1* in the *Boundary Selection* tab. At this point, put *r: T + Tx*. This step will add a *Neumann boundary condition*: $\dfrac{\partial \theta}{\partial X}(X = 0) = 0$, that is, $\dfrac{\partial T}{\partial x}(x = 0) = 0$, that is, $Tx = 0$ as given in Equation (3.24) (Note: as $T = r,\ r = T + Tx,\ Tx = 0$).

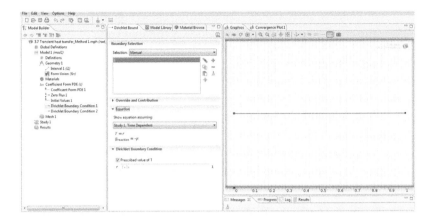

- Again right click on the *Coefficient Form PDE* tab in *Model Builder.* Select the *Dirichlet Boundary Condition* option again. Another window will open namely *Dirichlet Boundary Condition 2.* Click on it.
- Select the right point on the horizontal line graph. Click on the "+" sign at the top right corner, this will add *boundary 2* in the *Boundary Selection* tab. At this point, put *r: 0.* This step will add a *Dirichlet boundary condition:* $\theta = 0$ at $X = 1$, i.e., $T = 0$ at $x = 1$ as given in Equation (3.24) (Note: as $T = r$, $r = 0$, $T = 0$).

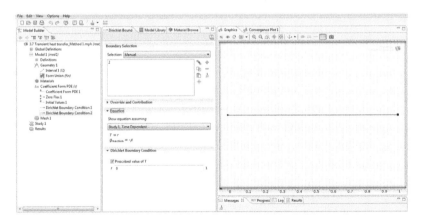

- Now, go to *Initial Values 1* available in the *Coefficient Form PDE* tab and set $T$ to 1. This will add the initial condition given in Equation (3.25).

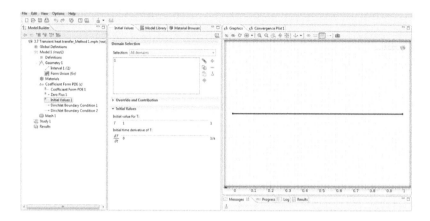

*Step 5*

- Click on the *Mesh* option in *Model Builder*. Select *Normal Mesh* Type. Click on *Build All* option at the top of ribbon. A dialogue box will appear in the *Message* tab as: *"Complete mesh consists of 15 elements."*

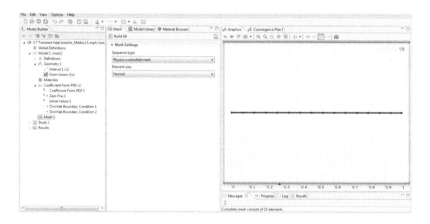

- Now, go to *Study* option in the model pellet tab. Click on the *Compute (=)* button. A graph will appear giving the profile of *Temperature versus x* as shown in Figure 3.17.
- Save the simulation.

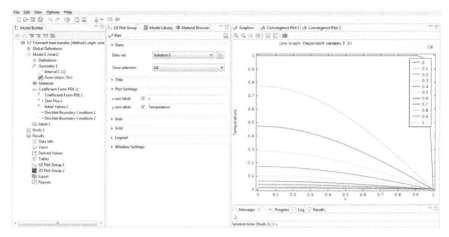

**FIGURE 3.17**  Solution of heat transfer problem [see Equation (3.22)].

*Step 6*

- In the *Model Builder*, right click on the *Results* tab and select the *2D Plot Group* option. Another window will open namely *2D Plot Group 2*. Right click on it and select *Surface*. Another window will open namely *Surface 1*. Then select *Height Expression* by right clicking on it. A 3D plot appears as shown in Figure 3.18.

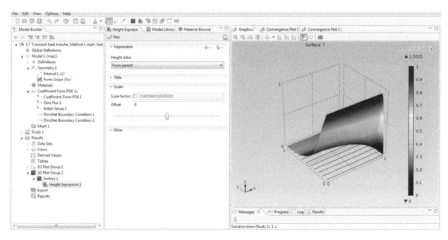

**FIGURE 3.18**  Solution of Equation (3.22) in 3D plot.

### 3.7.3   SIMULATION APPROACH (METHOD 2)

*Step 1*

- Open COMSOL Multiphysics.
- Select *1D Space Dimension* from the list of options. Hit the *next* arrow at the upper right corner.

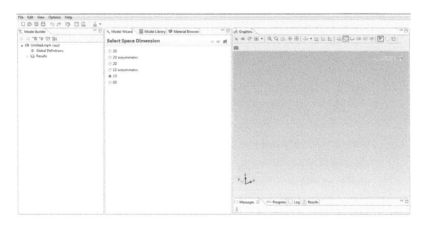

- Select and expand the *Heat Transfer* folder from the list of options in *Model Wizard*. Further select and *Heat Transfer in Solids*. Again hit the *next* arrow.

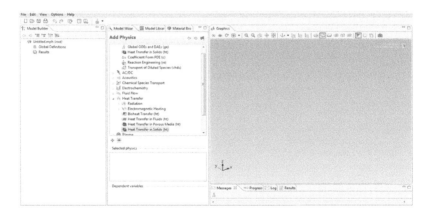

- Then, select *Time Dependent* from the list of *Study Type* options, and click on the *Finish flag* at the upper right corner of the application.

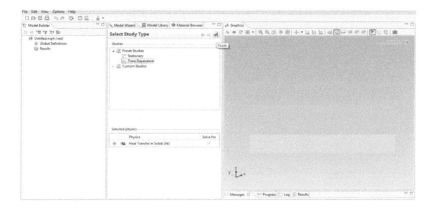

*Step 2*

- With the *Model Builder*, right click on the *Geometry* option. From the list of options, click on the *Interval* taskbar.
- In the *Interval* environment, select *Intervals*1, *Left endpoint: 0 and Right endpoint: 1*. Click on the *Build All* option at the upper part of tool bar. A straight line graph will appear on the right side of the application window.

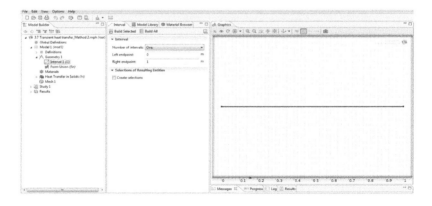

- Now in the *Model Builder*, select *Heat Transfer in Solids 1* option. This will open a tab to enter the coefficients of the characteristics of a model equation.
- In the *Domain Selection* panel, you will see an equation of the form

$$\tilde{n}C_p \frac{\partial T}{\partial t} + \tilde{n}C_p u.\nabla T = \nabla.(k\nabla T) + Q \qquad (3.27)$$

where $\nabla = \left[\dfrac{\partial}{\partial x}\right]$.

In order to solve the governing differential equation, we need to assign the coefficients in above equation a suitable value.

To convert Equation (3.27) to the desired form of Equation (3.22), the value of coefficients in Equation (3.27) to be changed as follows:

*k=2*

*ρ=1*

$C_p$*=1.*

This adjustment will reduce Equation (3.27) to Equation (3.22) format.

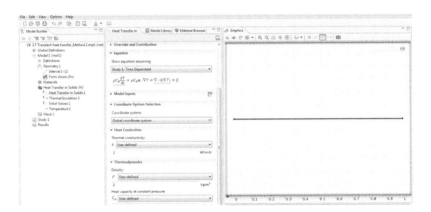

*Step 4*

- In the *Model Builder*, right click on the *Heat Transfer in Solids* tab and select *Temperature* option. Another window will open namely *Temperature 1*. Click on it.
- Select the left point on the horizontal line graph. Click on the "+" sign at the top right corner, this will add *boundary 2* in the *Boundary Selection* tab. At this point, put $T_0$: 0. This step will add a *Dirichlet*

*boundary condition*: $\theta = 0$ *at* $X = 1$, that is, $T = 0$ at $x = 1$ as given in Equation (3.24) (Note: as $T = r$, $r = 0$, $T = 0$).

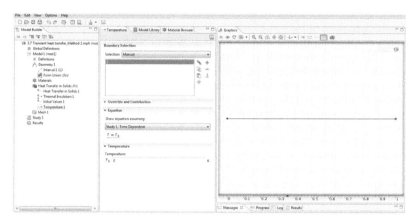

- Now, select *Thermal Insulation 1* option available in *Heat Transfer in Solids*. Select the left point on the horizontal line graph. The equation at this point is: $-n.(-k\nabla T)=0$. This is equivalent to a *Neumann boundary condition*: $\dfrac{\partial\theta}{\partial X}(X = 0) = 0$, that is, $\dfrac{\partial T}{\partial x}(x = 0) = 0$, that is, $Tx = 0$ as given in Equation (3.24)

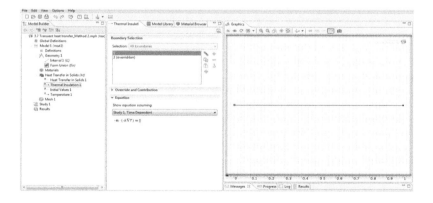

- Now, go to *Initial Values 1* available in the *Heat Transfer in Solids* tab and set $T$ to 1. This will add the initial condition given in Equation (3.25).

## Step 5

- Click on the *Mesh* option in *Model Builder.* Select *Normal Mesh* Type. Click on *Build All* option at the top of ribbon. A dialogue box will appear in the *Message* tab as: *"Complete mesh consists of 15 elements."*

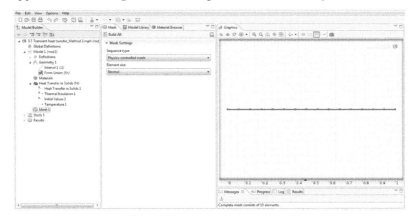

- Now, go to *Study* option in the model pellet tab. Click on the *Compute* (=) button. A graph will appear giving the profile of *Temperature versus x* as shown in Figure 3.19.
- Save the simulation.

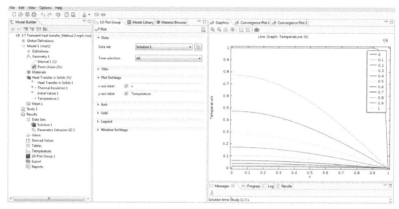

**FIGURE 3.19**    Solution of heat transfer problem [see Equation (3.22)].

*Step 6*

- In the *Model Builder*, right click on the *Results* tab and select the *2D Plot Group* option. Another window will open namely *2D Plot Group 2*. Right click on it and select *Surface*. Another window will open namely *Surface 1*. Then, select *Height Expression* by right clicking on it. A 3D plot appears as shown in Figure 3.20.

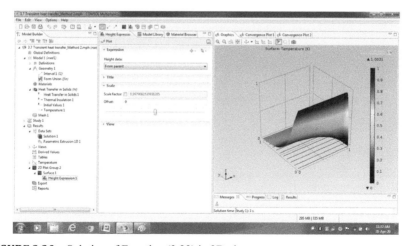

**FIGURE 3.20**    Solution of Equation (3.22) in 3D plot.

## 3.8   SIMULATION OF LINEAR ADSORPTION

### 3.8.1   *PROBLEM STATEMENT*

The linear adsorption is described by the following two differential equations (Finlayson, 1992; Rhee et al., 1986):

$$\frac{\partial c}{\partial t} + \frac{\partial c}{\partial x} + \frac{(1-\varnothing)}{\varnothing}(\gamma c - n) = \frac{1}{P_e}\frac{\partial^2 c}{\partial x^2}, \ \frac{1}{P_e} = \frac{\varnothing V^2}{Dk} \text{ or } P_e = \frac{Dk}{\varnothing V^2} \quad (3.28)$$

$$\frac{\partial n}{\partial t} = \gamma c - n. \quad (3.29)$$

Equation (3.28) represents the mass balance on the fluid phase with the following boundary and initial conditions:

*Boundary conditions:*

$$c(0,t') = 1 \quad (3.30)$$

$$D\frac{\partial c}{\partial x} = 0 \quad (3.31)$$

*Initial condition:*

$$c(x',0) = 0 \quad (3.32)$$

Equation (3.29) represents the mass balance of stationary phase with the following initial condition:

*Initial condition:*

$$n(x',0) = 0 \quad (3.33)$$

where
   $c$ = concentration of fluid
   $n$ = concentration on solid adsorbent
   $\varnothing$ = void fraction in the bed
   $V$ = fluid velocity
   $t'$ = time
   $x'$ = distance down the bed
   $k$ = mass transfer coefficient
   $\gamma$ = slope of the equilibrium line
   $D$ = diffusion coefficient
   $P_e$ = Peclet number.

### 3.8.2   SIMULATION APPROACH (METHOD 1)

*Step 1*

- Open COMSOL Multiphysics.
- Select *1D Space Dimension* from the list of options. Hit the *next* arrow at the upper right corner.

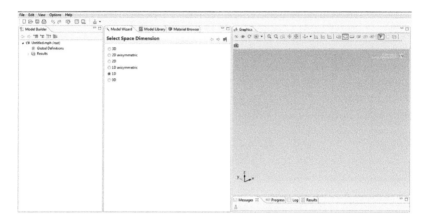

- Select and expand the *Mathematics* folder from the list of options in *Model Wizard*. Further select and expand *PDE Interface* and click on the *Coefficient Form PDE (c)*. Again hit the *next* arrow.

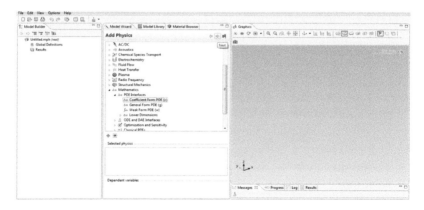

- Then select *Time Dependent* from the list of *Study Type* options, and click on the *Finish flag* at the upper right corner of the application.

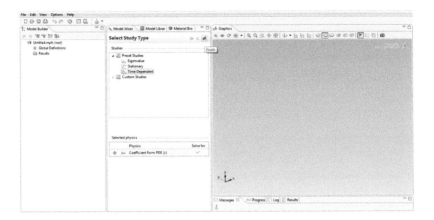

*Step 2*

- With the *Model Builder*, right click on the *Geometry* option. From the list of options, click on the *Interval* taskbar.
- In the *Interval* environment, select *Intervals* 1, *Left endpoint: 0 and Right endpoint: 1*. Click on the *Build All* option at the upper part of tool bar. A straight line graph will appear on the right side of the application window.

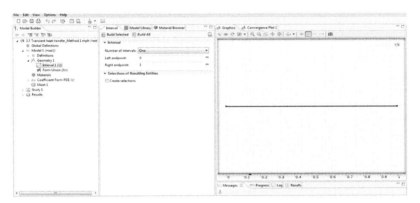

*Step 3*

- In the *Model Builder*, select *Coefficient Form PDE* node and change the Dependent variables to "*c*."

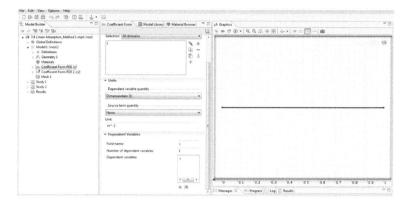

- Now, in the *Model Builder*, select *Coefficient Form PDE 1* option. This will open a tab to enter the coefficients of the characteristics of a model equation.
- In the *Domain Selection* panel, you will see an equation of the form

$$e_a \frac{\partial^2 c}{\partial t^2} + d_a \frac{\partial c}{\partial t} + \nabla \cdot \left( -c\nabla c - \alpha c + \gamma \right) + \beta \cdot \nabla c + ac = f \qquad (3.34)$$

where $\nabla = \left[ \dfrac{\partial}{\partial x} \right]$.

In order to solve the governing differential equation, we need to assign the coefficients in above equation a suitable value.

To convert Equation (3.34) to the desired form of Equation (3.28), the value of coefficients in Equation (3.34) to be changed as follows:

$c = 1/P_e$    $\alpha = 0,\ \beta = 1,\ \gamma = 0$

$a = 0$      $f = -\text{rate}$

$e_a = 0$      $d_a = 1$.

This adjustment will reduce Equation (3.34) to Equation (3.28) format.

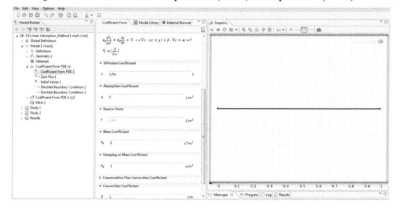

- In the *Model Builder*, right click on the *Definitions* tab and select *Variables*. Another window will open namely *Variables 1*. Click on it and define *raten: gamma\*c – n* and *rate: (1 – phi)\*raten/phi*.

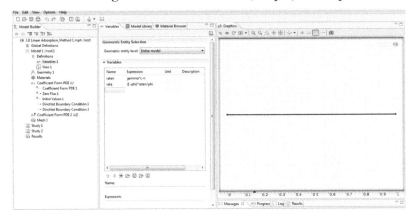

- In the *Model Builder*, right click on the *Global Definitions* tab and select *Parameters*. Another window will open namely *Parameters*. Click on it and define *phi: 0.4*, *gamma: 2*, and $P_e$: *1000*.

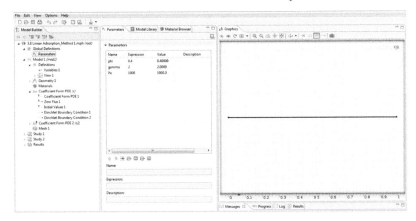

## Step 4

- Now, right click on the *Coefficient Form PDE* tab in *Model Builder*. Select the *Dirichlet Boundary Condition* option. Another window will open namely *Dirichlet Boundary Condition 1*. Click on it.
- Select the left point on the horizontal line graph. Click on the "+" sign at the top right corner, this will add *boundary 1* in the *Boundary*

*Selection* tab. At this point, put *r: 1*. This step will add a *Dirichlet boundary condition*: $c(0, t') = 1$ as given in Equation (3.30) (Note: as $c = r, r = 1, c = 1$).

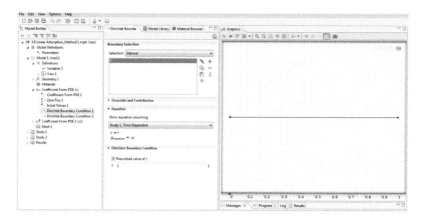

- Again right click on the *Coefficient Form PDE* tab in *Model Builder*. Select the *Dirichlet Boundary Condition* option again. Another window will open namely *Dirichlet Boundary Condition 2*. Click on it.
- Select the right point on the horizontal line graph. Click on the "+" sign at the top right corner, this will add *boundary 2* in the *Boundary Selection* tab. At this point, put *r: c + cx*. This step will add a *Neumann boundary condition*: $D\dfrac{\partial c}{\partial x} = 0$, that is, $\dfrac{\partial c}{\partial x} = 0$, that is, $cx = 0$, as given in Equation (3.31) (Note: as $c = r, r = c + cx, cx = 0$).

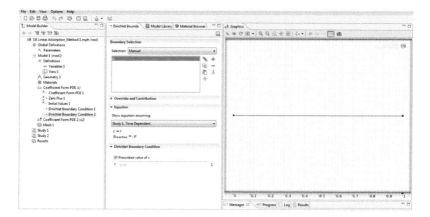

- Now, go to *Initial Values 1* available in the *Coefficient Form PDE* tab and set *c* to 0. This will add the initial condition given in Equation (3.32).

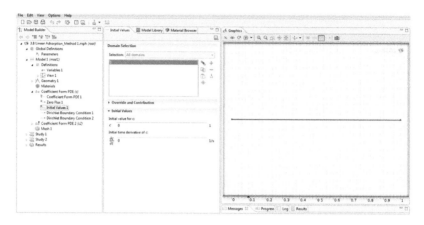

## Step 5

- To add Equation (3.29), right click on the *Model 1* tab and select *Add Physics*. Expand the *Mathematics* folder from the list of options in *Model Wizard*. Further select and expand *PDE Interface* and click on the *Coefficient Form PDE (c)*. Again hit the *next* arrow.
- Then, select *Time Dependent* from the list of *Study Type* options, and click on the *Finish flag* at the upper right corner of the application.

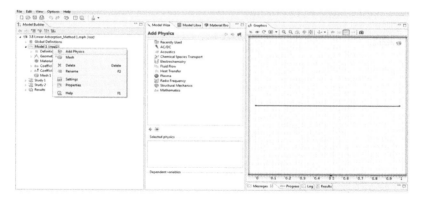

## Step 6

- In the *Model Builder*, select *Coefficient Form PDE 2* node and change the Dependent variables to *n*.

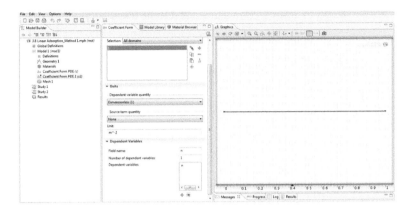

- Now, in the *Model Builder*, select *Coefficient Form PDE 1* available in *Coefficient Form PDE 2* option. This will open a tab to enter the coefficients of the characteristics of a model equation.
- In the *Domain Selection* panel, you will see an equation of the form

$$e_a \frac{\partial^2 n}{\partial t^2} + d_a \frac{\partial n}{\partial t} + \nabla \cdot \left(-c\nabla n - \alpha n + \gamma\right) + \beta \cdot \nabla n + an = f \qquad (3.35)$$

where $\nabla = \left[\dfrac{\partial}{\partial x}\right]$.

In order to solve the governing differential equation, we need to assign the coefficients in above equation a suitable value.

To convert Equation (3.35) to the desired form of Equation (3.29), the value of coefficients in Equation (3.35) to be changed as follows:

$c = 0\ \alpha = 0,\ \beta = 0,\ \gamma = 0$

$a = 0\ f = raten$

$e_a = 0\ d_a = 1.$

This adjustment will reduce Equation (3.35) to Equation (3.29) format.

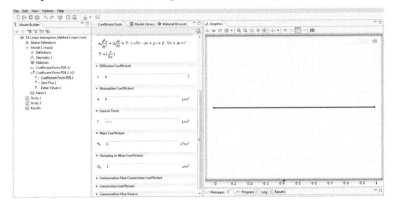

- Now, go to *Initial Values 1* available in the *Coefficient Form PDE 2* tab and set *n* to 0. This will add the initial condition given in Equation (3.33).

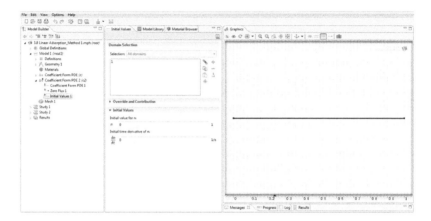

*Step 7*

- Click on the *Mesh* option in *Model Builder*. Select *Extrafine Mesh* Type. Click on *Build All* option at the top of ribbon. A dialogue box will appear in the *Message* tab as: *"Complete mesh consists of 50 elements."*

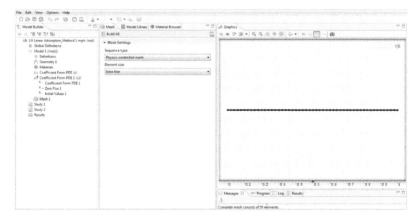

- Now, go to *Study 1* option in the model pellet tab. Click on the *Compute (=)* button. A graph will appear giving the profile of *c* versus *x* as shown in Figure 3.21.
- Save the simulation.

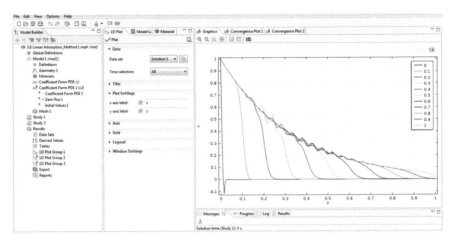

**FIGURE 3.21**   Concentration in the fluid.

### 3.8.3   *SIMULATION APPROACH (METHOD 2)*

*Step 1*

- Open COMSOL Multiphysics.
- Select *1D Space Dimension* from the list of options. Hit the *next* arrow at the upper right corner.

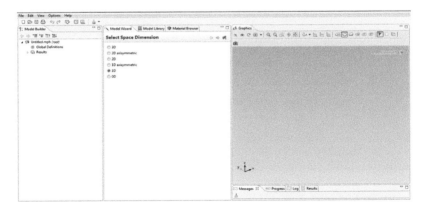

- Expand the *Chemical Species Transport* folder from the list of options in *Model Wizard* and select *Transport of Diluted Species*. Hit the *next* arrow again.

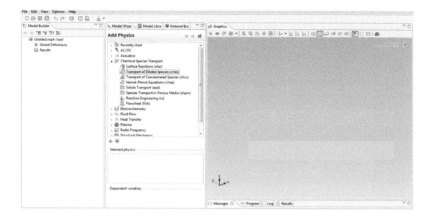

- Next, select *Time Dependent* from the list of *Study Type* options, and click on the *Finish flag* at the upper right corner of the application.

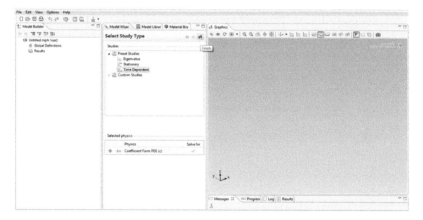

*Step 2*

- With the *Model Builder*, right click on the *Geometry* option. From the list of options, click on the *Interval* taskbar.
- In the *Interval* environment, select *Intervals*1, *Left endpoint: 0 and Right endpoint: 1*. Click on the *Build All* option at the upper part of tool bar. A straight line graph will appear on the right side of the application window.

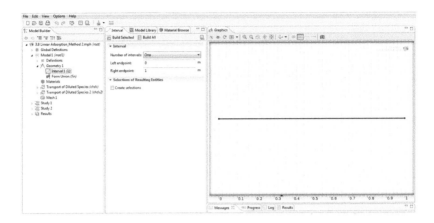

*Step 3*

- In the *Model Builder*, select the *Transport of Diluted Species* tab and expand the Dependent variables tab. Change the species concentration to "*c*."

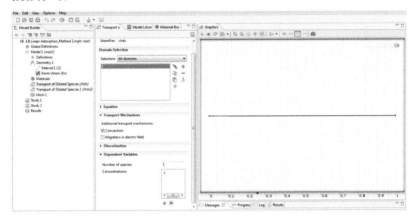

- Now, in the *Model Builder*, select *Transport of Diluted Species* option. This will open a tab to enter the coefficients of the characteristics of a model equation.
- In the *Domain Selection* panel, you will see an equation of the form

$$\frac{\partial c_i}{\partial t} + \nabla \cdot \left(-D_i \nabla c_i\right) + u \cdot \nabla c_i = R_i \tag{3.36}$$

$$N_i = -D_i \nabla c_i + u c_i.$$

In order to solve the governing differential equation, we need to assign the coefficients in above equation a suitable value.

To convert Equation (3.36) to the desired form of Equation (3.28), the value of coefficients in Equation (3.36) to be changed as follows: Select *Convection and Diffusion 1* tab and set *u: 1.0* and *Dc: 1/P$_e$*.

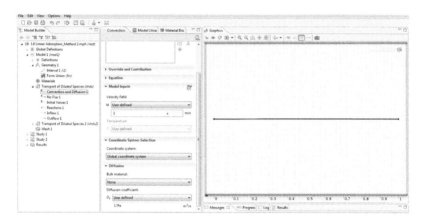

- Then, right click *Transport of Diluted Species* option and choose *Reactions*. This will open the new window *Reactions 1*, set *Rc* to – *rate*. Select the horizontal line graph. Click on the "+" sign at the top right corner, this will add *Domain 1* in the *Domain Selection* tab.

- In the *Model Builder*, right click on the *Definitions* tab and select *Variables*. Another window will open namely *Variables 1*. Click on it and define *raten: gamma\*c – n* and *rate: (1 – phi)\*raten/phi*.

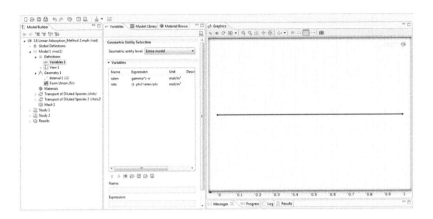

- In the *Model Builder*, right click on the *Global Definitions* tab and select *Parameters*. Another window will open namely *Parameters*. Click on it and define *phi: 0.4, gamma: 2,* and $P_e$: *1000*.

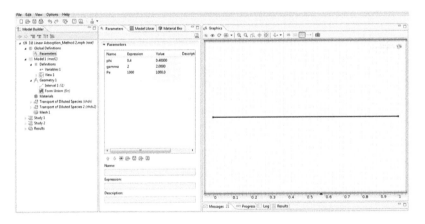

## Step 4

- Now, right click on the *Transport of Diluted Species* tab in *Model Builder*. Select the *Inflow* option. Another window will open namely *Inflow 1*. Click on it.
- Select the left point on the horizontal line graph. Click on the "+" sign at the top right corner; this will add *boundary 1* in the *Boundary Selection* tab. At this point, put $c_{0,c}$: *1*. This step will add a *Dirichlet boundary condition*: $c(0, t') = 1$ as given in Equation (3.30).

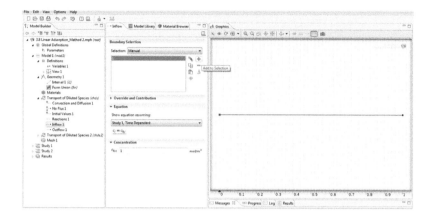

- Again right click on the *Transport of Diluted Species* tab in *Model Builder.* Select the *Outflow* option. Another window will open namely *Outflow 1.* Click on it.

- Select the right point on the horizontal line graph. Click on the "+" sign at the top right corner; this will add *boundary 2* in the *Boundary Selection* tab. This will add the equation $-n.D_i \nabla_{ci} = 0$, which is same as the *Neumann boundary condition*: $D\dfrac{\partial c}{\partial x} = 0$ as given in Equation (3.31).

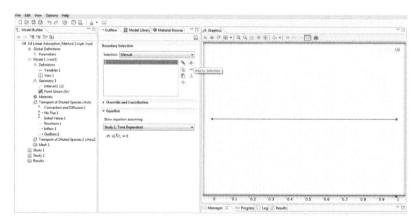

- Now, go to *Initial Values 1* available in the *Transport of Diluted Species* tab and set $c$ to 0. This will add the initial condition given in Equation (3.32).

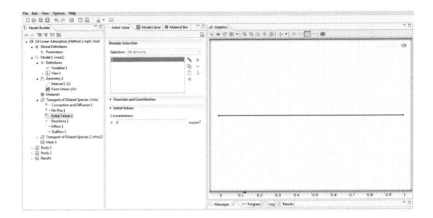

## Step 5

- To add Equation (3.29), right click on the *Model 1* tab and select *Add Physics*. Expand the *Mathematics* folder from the list of options in *Model Wizard*. Further select and expand *PDE Interface* and click on the *Coefficient Form PDE (c)*. Again hit the *next* arrow.
- Next, select *Time Dependent* from the list of *Study Type* options, and click on the *Finish flag* at the upper right corner of the application.

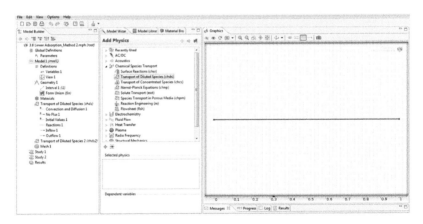

## Step 6

- In the *Model Builder*, select the *Transport of Diluted Species 2* tab and expand the Dependent variables tab. Change the species concentration to *"n."*

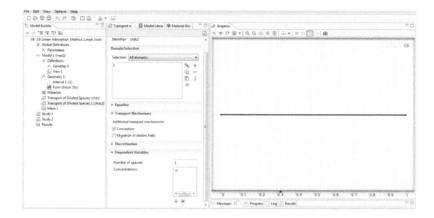

- Now, in the *Model Builder*, select *Transport of Diluted Species* option. This will open a tab to enter the coefficients of the characteristics of a model equation.
- In the *Domain Selection* panel, you will see an equation of the form

$$\frac{\partial c_i}{\partial t} + \nabla \cdot \left(-D_i \nabla c_i\right) + u \cdot \nabla c_i = R_i \tag{3.37}$$

$$N_i = -D_i \nabla c_i + u c_i.$$

In order to solve the governing differential equation, we need to assign the coefficients in above equation a suitable value.

To convert Equation (3.36) to the desired form of Equation (3.29), the value of coefficients in Equation (3.37) to be changed as follows: Select *Convection and Diffusion 1* tab and define *u: 0* and *Dn: 0*.

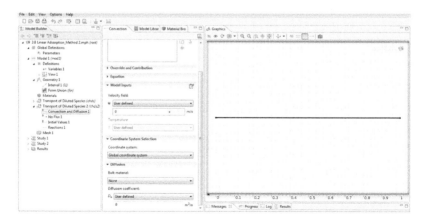

- Then, select *Reactions* by right clicking on *Transport of Diluted Species 2* option. This will open the new window *Reactions 1*, set *Rn* to *raten*. Select the horizontal line graph. Click on the "+" sign at the top right corner; this will add *Domain 1* in the *Domain Selection* tab.

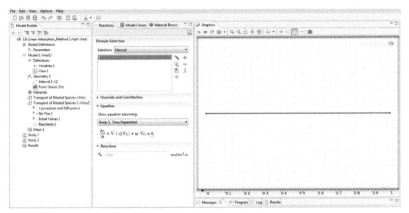

- Now, go to *Initial Values 1* tab and set *n:* 0. This will add the initial condition given in Equation (3.33).

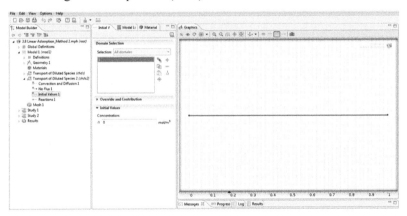

*Step 7*

- Click on the *Mesh* option in *Model Builder.* Select *Extrafine Mesh* Type. Click on *Build All* option at the top of ribbon. A dialogue box will appear in the *Message* tab as: *"Complete mesh consists of 50 elements."*

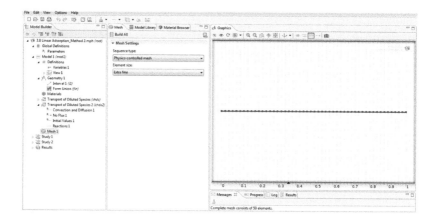

- Now, go to *Study 1* option in the model pellet tab. Click on the *Compute* (=) button. A graph will appear giving the profile of *c versus x*.
- Save the simulation.

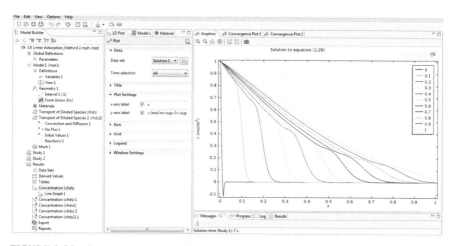

**FIGURE 3.22** Concentration in the fluid.

## 3.9 SIMULATION OF AN INFINITELY LONG CYLINDER

### 3.9.1 *PROBLEM STATEMENT*

An infinitely long cylinder of radius $r_0$ and temperature $T_i$ described by the following differential equation. It is immersed in a bath of hot fluid maintained at $T_\infty$. The heat transfer coefficient of the bath and the cylindrical surface is $h$. Assume the constant $k$, $\rho$, $C$ of the cylinder.

The governing differential equation is given by

$$\frac{\partial \theta}{\partial \tau} = \frac{\partial^2 \theta}{\partial R^2} + \frac{1}{R}\frac{\partial \theta}{\partial R} \qquad (3.38)$$

where,

$$\theta = \frac{T - T_\infty}{T_\infty - T_i}, \quad R = \frac{r}{r_0}, \quad \tau = \frac{\alpha t}{r_0^2} \qquad (3.39)$$

$\theta$, $R$ and $\tau$ are dimensionless temperature, dimensionless radius, and dimensionless time, respectively.

Initial and boundary conditions:

$$\text{I.C.:} \quad \tau = 0, \qquad \theta = 0 \qquad (3.40)$$

$$\text{B.C.1:} \ R = 1, \qquad H\frac{\partial \theta}{\partial R} = 1 - \theta \qquad (3.41)$$

$$\text{B.C.2:} \ R = 0, \qquad \frac{\partial \theta}{\partial R} = 0 \qquad (3.42)$$

Discretize the above equations to find the temperature history $T(r, t)$ till 1 s with a step of 0.1 s. Take $H = 2$.

### 3.9.1 SIMULATION APPROACH

*Step 1*

- Open COMSOL Multiphysics.
- Select *1D* axisymmetric *Space Dimension* from the list of options. Hit the *next* arrow at the upper right corner.

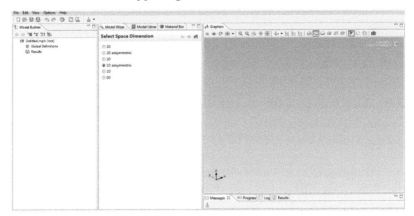

- Select and expand the *Mathematics* folder from the list of options in *Model Wizard*. Further select and expand *PDE Interface* and click on the *Coefficient Form PDE (c)*. Again hit the *next* arrow.

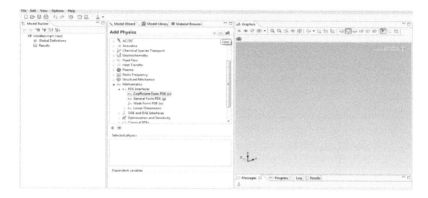

- Then select *Time Dependent* from the list of *Study Type* options, and click on the *Finish flag* at the upper right corner of the application.

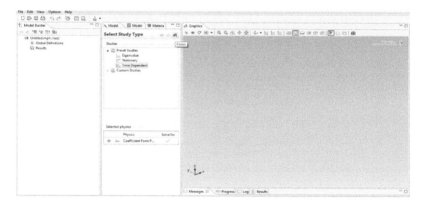

*Step 2*

- With the *Model Builder*, right click on the *Geometry* option. From the list of options, click on the *Interval* taskbar.
- In the *Interval* environment, select *Intervals* 1, *Left endpoint: 0 and Right endpoint: 1*. Click on the *Build All* option at the upper part of tool bar. A straight line graph will appear on the right side of the application window.

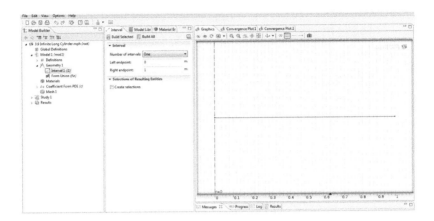

### Step 3

- In the *Model Builder*, select *Coefficient Form PDE* node and change the Dependent variables to "*Theta*".

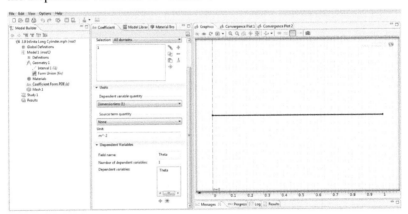

- Now, in the *Model Builder*, select *Coefficient Form PDE 1* option. This will open a tab to enter the coefficients of the characteristics of a model equation.
- In the *Domain Selection* panel, you will see an equation of the form

$$e_a \frac{\partial^2 \text{Theta}}{\partial t^2} + d_a \frac{\partial \text{Theta}}{\partial t} + \nabla \cdot \left( -c\nabla \text{Theta} - \alpha \text{Theta} + \gamma \right)$$
$$+ \beta \cdot \nabla \text{Theta} + a\text{Theta} = f \tag{3.43}$$

where $\nabla = \left[ \dfrac{\partial}{\partial r} \right]$.

In order to solve the governing differential equation, we need to assign the coefficients in above equation a suitable value.

To convert Equation (3.43) to the desired form of Equation (3.38), the value of coefficients in Equation (3.43) to be changed as follows:

$c = 1$ $\alpha = 0$, $\beta = 0$, $\gamma = 0$

$a = 0$ $f = Thetar/r$

$e_a = 0$ $d_a = 1$.

This adjustment will reduce Equation (3.43) to Equation (3.38) format.

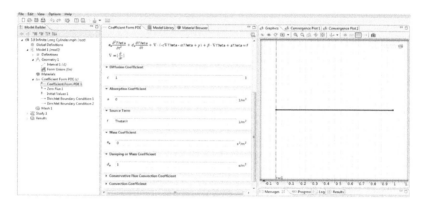

*Step 4*

- Now, right click on the *Coefficient Form PDE* tab in *Model Builder*. Select the *Dirichlet Boundary Condition* option. Another window will open namely *Dirichlet Boundary Condition 1*. Click on it.

- Select the left point on the horizontal line graph. Click on the "+" sign at the top right corner, this will add *boundary 1* in the *Boundary Selection* tab. At this point, put $r$: Theta + Thetar. This step will add the *Neumann Boundary Condition* $\dfrac{d\theta}{dR} = 0$, that is, $\dfrac{\partial \text{Theta}}{\partial r} = 0$, that is, *Thetar* = 0 as given in Equation (3.42) (Note: as Theta = r, r = Theta + Thetar; Thetar = 0).

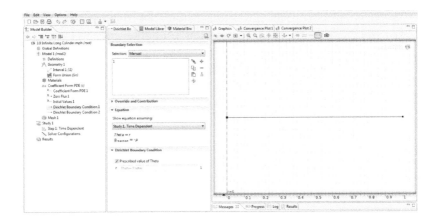

- Again right click on the *Coefficient Form PDE* tab in *Model Builder.* Select the *Dirichlet Boundary Condition* option again. Another window will open namely *Dirichlet Boundary Condition 2.* Click on it.

- Select the right point on the horizontal line graph. Click on the "+" sign at the top right corner, this will add *boundary 2* in the *Boundary Selection* tab. At this point, put *r:* Theta + Theta*r* – *(1 – Theta)/2.* This step will add the *Neumann Boundary Condition*

$$H \frac{d\theta}{dR} = 1 - \theta, \text{that is, } H \frac{\partial \text{Theta}}{\partial r} = 1 - \text{Theta, that is Theta}r = (1 - \text{Theta})/2$$

if *H = 2* as given in Equation (3.41) [Note: as Theta = *r, r* = Theta + Theta*r* – *(1* – Theta)/2, Theta*r* = *(1* – Theta)/2].

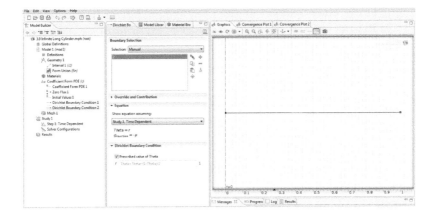

- Now, go to *Initial Values 1* available in the *Coefficient Form PDE* tab and set *Theta* to 0. This will add the initial condition given in Equation (3.40).

*Step 5*

- Click on the *Mesh* option in *Model Builder.* Select *Normal Mesh* Type. Click on *Build All* option at the top of ribbon. A dialogue box will appear in the *Message* tab as: *"Complete mesh consists of 15 elements."*

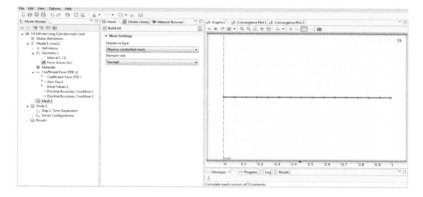

- Now, go to *Study* option in the model pellet tab. Click on the *Compute (=)* button. A graph will appear giving the profile of *Theta versus r* as shown in Figure 3.23.
- Save the simulation.

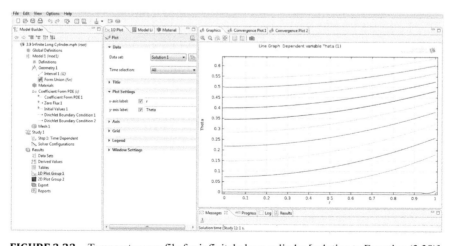

**FIGURE 3.23**  Temperature profile for infinitely long cylinder [solution to Equation (3.38)].

## 3.10　PROBLEMS

1. Consider the differential equation where reaction and diffusion are taking place in a pore of length 1 mm

$$\frac{d^2C}{dx^2} - \frac{k}{D}C = 0$$

The boundary conditions are

$$\text{At } x = 0, C = C_s$$

At $x = L, \dfrac{dC}{dx} = 0$

$C_s = 1$ mol/m³, concentration at the surface of the pore
$k = 10^{-3}$ s⁻¹, rate constant
$D = 10^{-9}$ m²/s, effective diffusivity.
Determine the concentration at $x = 0.5$ mm. Divide the pore length into 100 parts.

2. Consider the following convection and diffusion equation:

$$-u\frac{dC}{dx} + D\frac{d^2C}{dx^2} = 0.$$

The boundary conditions are
*At x = 0, C = 1*
At $x = 1$, m, $C = 0$.
Determine the concentration $C$. Make 10 parts between $x = 0$ and $x = 1$ m.

3. Consider a reaction $A \rightarrow B$ carried out in a tubular reactor of length 10 m

$$D\frac{d^2C_A}{dx^2} - u\frac{dC_A}{dx} - kC_A = 0.$$

The boundary conditions are

At $x = 0(\text{inlet}), uC_{A,\text{in}} = uC_A - D\dfrac{dC_A}{dx}$

At $x = 10\text{m (exit)}, \dfrac{dC_A}{dx} = 0$

$u = 1$ m/s, axial velocity of fluid $A$
$D = 10^{-4}$ m²/s, axial dispersion coefficient,
$k = 0.10$ s⁻¹, rate constant of the reaction

$C_{A'\text{in}} = 1$ mol/m³, inlet concentration. Determine the concentration of $A$. Make 50, 20, and 10 parts of the reactor.

4. Consider the following reaction and diffusion equation of a catalyst pellet with radius of 1 cm

$$D\frac{d^2C_A}{dr^2} + \frac{2}{r}D\frac{dC_A}{dr} - kC_A = 0.$$

The boundary conditions are

At $r = R$, $C_A = 1$

At $r = 0, \dfrac{dC_A}{dr} = 1$

$D = 10^{-9}$ m²/s, effective binary diffusivity of $A$

$k = 0.1$ s⁻¹, rate constant of the reaction

$C_A = 1$ mol/m³, concentration at the surface of the spherical catalyst pellet.

Determine the concentration of $A$ along the radius of the catalyst pellet.

5. Consider the following reaction and diffusion equation in a nonisothermal spherical catalyst pellet of radius 1 cm

$$\frac{d^2C}{dr^2} + \frac{2}{r}\frac{dC}{dr} - \frac{k}{D}C = 0$$

where

$$k(T) = k(T_s)\exp\left[-\frac{\gamma\beta(C - C_s)}{1 - \beta(C - C_s)}\right].$$

The boundary conditions are

At $r = 1$ cm, $C = 1$

At $r = 0, \dfrac{dC}{dr} = 0$

$D = 10^{-9}$ m²/s, effective diffusivity

$k = (T_s) = 10^{-3}$ s⁻¹, rate constant of the reaction

$C_S = 1$ mol/m³, concentration at the surface of the spherical catalyst pellet

$\beta = 1$

$\gamma = 1$.

Determine the concentration along the radius of the catalyst pellet by making 100 parts.

6.  Consider the axial dispersion in a chemical reactor given by following differential equation:

$$\frac{1}{Pe}\frac{d^2c}{dx^2} - \frac{dc}{dx} - Da\frac{c}{c+v} = 0.$$

The boundary conditions are

$$-\frac{1}{Pe}\frac{dc}{dx}(0) = 1 - c(0)$$

$$\frac{dc}{dx}(1) = 0$$

Solve for $Da = 8$, $n = 3$, $Pe = 15,150,1500$.

7.  Consider the following reaction and diffusion equation in a packed bed:

$$\frac{\partial c}{\partial t} + \frac{dc}{dx} = \frac{1}{Pe}\frac{d^2c}{dx^2} - Da\frac{c}{c+v}.$$

The initial and boundary conditions are

$c(x, 0) = 0$

$$-\frac{1}{Pe}\frac{dc}{dx}(0,t) = 1 - c(0,t)$$

$$\frac{dc}{dx}(1,t) = 0$$

$Pe = 100$, $0 \le x \le 1$, $Da = 2$, $v = 2$.

# CHAPTER 4

# Fluid Flow

## 4.1 INTRODUCTION

In this chapter, the application of *COMSOL* is discussed to solve problems in fluid flow using *Mathematics* and *Chemical Engineering* modules. The examples discussed here include entry flow in a pipe considering a Newtonian and non-Newtonian fluid, application on microfluidic devices, turbulent flow, start-up flow, and flow through an orifice. The examples demonstrate the application of periodic boundary conditions, making different plots such as streamlines, velocity, etc. to calculate different properties and application of parametric solver.

## 4.2 SIMULATION OF ENTRY FLOW OF A NEWTONIAN FLUID IN A PIPE

### 4.2.1 PROBLEM STATEMENT

The entry flow of a Newtonian fluid into a pipe as shown in Figure 4.1 is described mathematically by Navier–Stokes equation (4.1) and continuity equation (4.2) in the dimensionless form. An order of magnitude analysis reveals (Schlichting, 1979) that $z$-momentum in 2D boundary layer flow can be neglected. Therefore, in Cartesian coordinates, 2D, laminar, incompressible flow with constant viscosity is described by

$$\rho\left(u\frac{\partial u}{\partial r} + w\frac{\partial u}{\partial z}\right) = -\frac{\partial p}{\partial r} + \mu\left(\frac{\partial^2 u}{\partial r^2} + \frac{\partial^2 u}{\partial z^2}\right) \tag{4.1}$$

$$\frac{\partial u}{\partial r} + \frac{\partial w}{\partial z} = 0. \tag{4.2}$$

Boundary conditions:

At $z = 0$, $u = w = 1$ at any $r$ \hfill (4.3)

At $z = 2$, $p_0 = 0$ at any $r$.                                                    (4.4)

By default, the centerline has slip condition with velocity zero for $r$ component and nonzero or floating for $z$ component; hence, we have the following:

At $r = 0$, $u = 0$, $w$ is nonzero or floating at any $z$.                          (4.5)

For the wall at other boundary is having no slip condition with the velocity zero for both components, hence

At $r = 0.5$, $u = 0$; $w = 0$ at any $z$.                                           (4.6)

Given: $\rho = 10$, $\mu = 1$

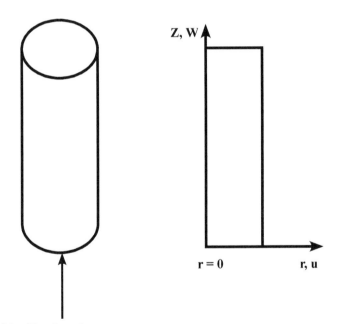

**FIGURE 4.1**   Flow in a pipe.

## 4.2.2   *SIMULATION APPROACH*

*Step 1*

- Open COMSOL Multiphysics.
- Select *2D* axisymmetric *Space Dimension* from the list of options. Hit the *next* arrow at the upper right corner.

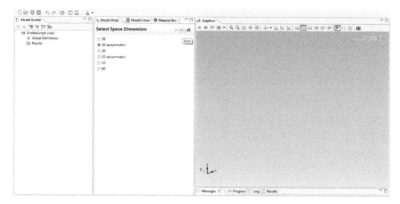

- Select and expand the *Fluid Flow* folder from the list of options in *Model Wizard* and select the *Laminar Flow*. Again hit the *next* arrow at the upper right corner.

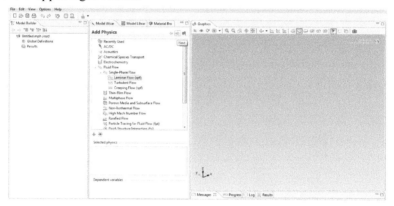

- Then, select *Stationary* from the list of *Study Type* options, and click on the *Finish flag* at the upper right corner of the application.

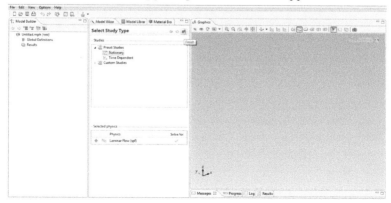

- Select the *Root* node and change the *Unit System: None*.

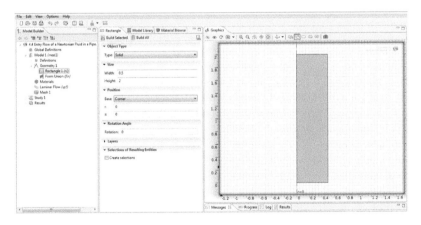

## Step 2

- Right click on the *Geometry* option and select the *Rectangle*.
- In the *Rectangle* environment, select *Size* with *Width: 0.5 and Height: 2*. Click on the *Build All* option at the upper part of tool bar. A rectangle graph will appear on the right side of the application window.

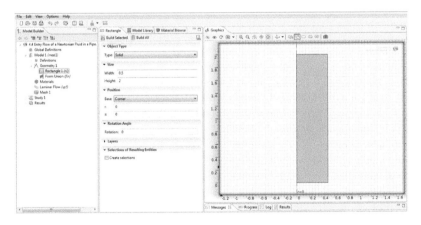

## Step 3

- In the *Laminar Flow* node, select the *Compressibility:Incompressible Flow*.

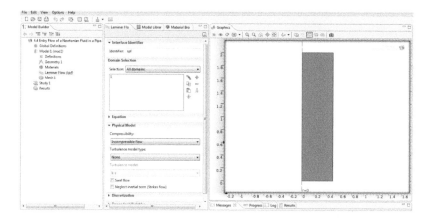

- Now, in the *Model Builder*, select *Fluid Properties 1* option available in *Laminar Flow*. This will open a tab to enter the coefficients of the characteristics of a model equation.
- In the *Domain Selection* panel, you will see an equation of the form

$$\rho\left(u.\nabla\right)u = \nabla.\left[-pl + \mu\left(\nabla u + \left(\nabla u\right)^{T}\right)\right] + F \qquad (4.7)$$

$$\rho\nabla.u = 0. \qquad (4.8)$$

To convert Equations (4.7) and (4.8) to the desired form of Equations (4.1) and (4.2), respectively, the value of coefficients in Equations (4.7) and (4.8) to be changed as follows:

$$\rho = 10, \mu = 1.$$

Then, click on the "+" sign at the top right corner; this will add *Domain 1* in the *Domain Selection* tab.

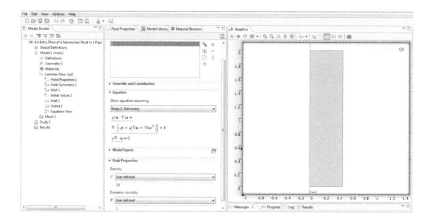

*Step 4*

- Now, right click on the *Laminar Flow* tab in *Model Builder.* Select the *Inlet* option. Another window will open namely *Inlet 1.* Click on it.
- Select the bottom boundary on the rectangle graph. Click on the "+" sign at the top right corner; this will add *boundary 2* in the *Boundary Selection* tab. At this point, put *Normal inflow velocity $U_0$: 1.* This step will add a boundary condition as given in Equation (4.3).

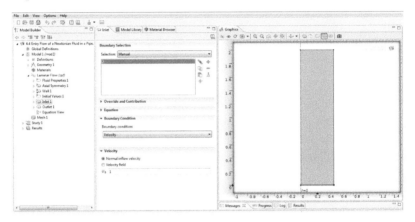

- Again right click on the *Laminar Flow* tab in *Model Builder.* Select the *Outlet* option. Another window will open namely *Outlet 1.* Click on it.
- Select the top boundary on the rectangle graph. Click on the "+" sign at the top right corner; this will add *boundary 3* in the *Boundary Selection* tab. At this point, put *Pressure $p_0$: 0.* This step will add a boundary condition as given in Equation (4.4).

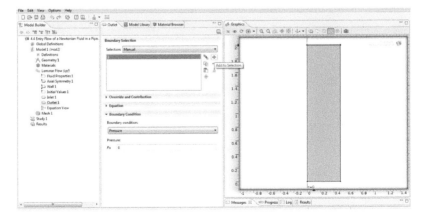

- Now, in the *Model Builder*, select *Wall 1* option available in *Laminar Flow*. By default, it has *boundary 4* in the *Boundary Selection* tab. At this point, put boundary condition to *No slip.* This step will add a boundary condition as given in Equation (4.6).

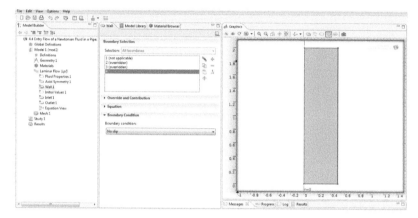

- Now, in the *Model Builder*, select *Axial Symmetry 1* option available in *Laminar Flow*. By default, it has *boundary 1* in the *Boundary Selection* tab. In COMSOL, by default, the axisymmetric geometry allows slip and this step will add a boundary condition as given in Equation (4.5).

*Step 5*

- Click on the *Mesh* option in *Model Builder.* Select *Normal Mesh* Type. Click on *Build All* option at the top of ribbon. A dialogue box will appear in the *Message* tab as: *"Complete mesh consists of 3353 elements."*

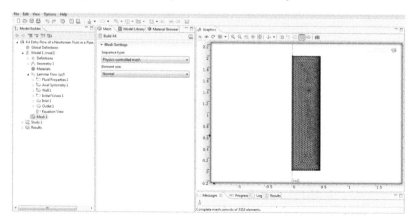

- Now, go to *Study 1* option in the model pellet tab. Click on the *Compute (=)* button. A plot of velocity magnitude, $U = sqrt(u2 + w2)$ will appear.
- Save the simulation.

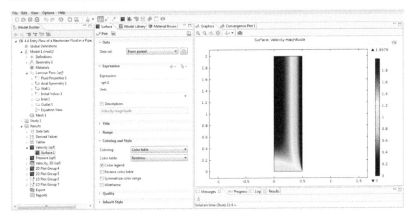

*Step 6*

- To plot the inlet velocity, click on the *Line Graph 1*. Another *Plot* window will open. Click on the "+" sign in the *y*-axis data and select *Velocity magnitude*, U.
- Select the bottom boundary on the rectangle graph. Click on the "+" sign at the top right corner, this will add *boundary 2* in the *Boundary Selection* tab. Click on the "+" sign in the *x*-axis data and select *r-coordinate (r)*. Plot the *Line Graph*. A graph will appear giving the profile of *Velocity magnitude versus r* for boundary 2.

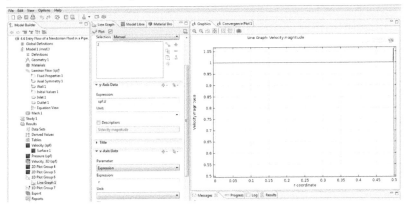

- Now, select the top boundary on the rectangle graph. Click on the "+" sign at the top right corner, this will add *boundary 3* in the *Boundary*

*Selection* tab. Click on the "+" sign in the *x*-axis data and select *r-coordinate (r)*. Plot the *Line Graph*. A graph will appear giving the profile of *Velocity magnitude versus r* for boundary 3 as shown in Figure 4.2.

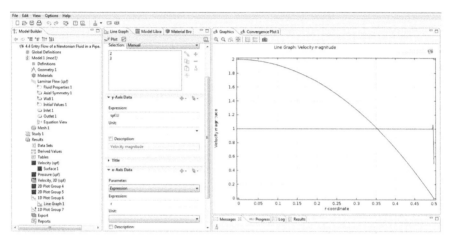

**FIGURE 4.2**    Velocity at inlet and outlet.

- Now, select the left boundary on the rectangle graph. Click on the "+" sign at the top right corner, this will add *boundary 1* in the *Boundary Selection* tab. Click on the "+" sign in the *x*-axis data and select *z-coordinate (z)*. Plot the *Line Graph*. A graph will appear giving the profile of *Velocity magnitude versus z* for boundary 1 as shown in Figure 4.3.

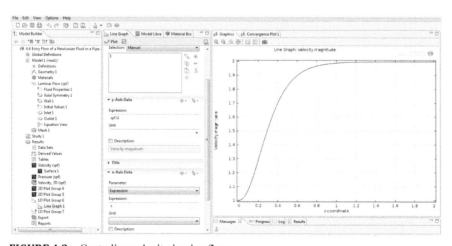

**FIGURE 4.3**    Centerline velocity in pipe flow.

## 4.3    SIMULATION OF ENTRY FLOW OF A NON-NEWTONIAN FLUID IN A PIPE

### 4.3.1    PROBLEM STATEMENT

The entry flow of a non-Newtonian fluid into a pipe as shown in Figure 4.1 is described mathematically by Navier–Stokes equation (4.9) and continuity equation (4.10) in the dimensionless form. An order of magnitude analysis reveals (Schlichting, 1979) that $z$-momentum in 2D boundary layer flow can be neglected. Therefore, in Cartesian coordinates, 2D, laminar, incompressible flow with changing shear viscosity is described by

$$\rho\left( u\frac{\partial u}{\partial r}+w\frac{\partial u}{\partial z}\right)=-\frac{\partial p}{\partial r}+\mu\left(\frac{\partial^2 u}{\partial r^2}+\frac{\partial^2 u}{\partial z^2}\right) \tag{4.9}$$

$$\frac{\partial u}{\partial r}+\frac{\partial w}{\partial z}=0. \tag{4.10}$$

Boundary conditions:

At $z = 0$, $u = w = 1$ at any $r$ $\hspace{3cm}$ (4.11)

At $z = 2$, $p_0 = 0$ at any $r$. $\hspace{3cm}$ (4.12)

By default, the centerline has slip condition with velocity zero for $r$ component and nonzero or floating for $z$ component; hence, we have the following:

At $r = 0$, $u = 0$, $w$ is nonzero or floating at any $z$. $\hspace{1cm}$ (4.13)

For the wall at other boundary is having no slip condition with the velocity zero for both components, hence

At $r = 0.5$, $u = 0$; $w = 0$ at any $z$. $\hspace{2.5cm}$ (4.14)

The shear viscosity is given by

$$\frac{\eta-\eta_\infty}{\eta_0-\eta_\infty}=\left[1+\left(\lambda\gamma\right)^2\right]^{(n-1)/2} \tag{4.15}$$

where

$\eta_\infty = 0.05$, $\eta_0 = 0.492$, $\lambda = 0.1$, $n = 0.4$, given $\rho = 10$.

### 4.3.2    SIMULATION APPROACH

*Step 1*

- Open COMSOL Multiphysics.
- Select *2D* axisymmetric *Space Dimension* from the list of options. Hit the *next* arrow at the upper right corner.

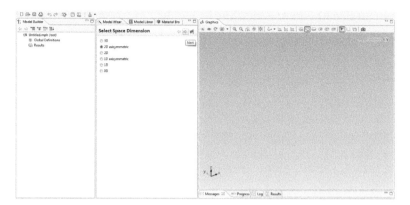

- Select and expand the *Fluid Flow* folder from the list of options in *Model Wizard* and select the *Laminar Flow*. Again hit the *next* arrow at the upper right corner.

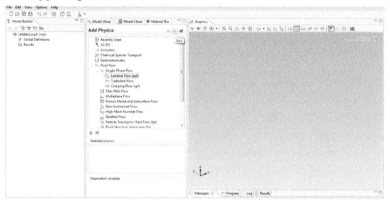

- Then, select *Stationary* from the list of *Study Type* options, and click on the *Finish flag* at the upper right corner of the application.

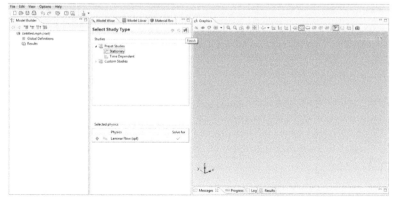

- Select the *Root* node and change the *Unit System: None.*

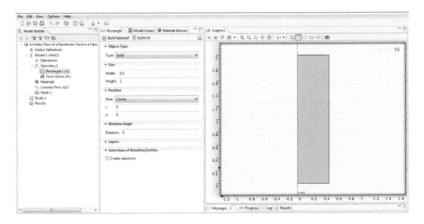

*Step 2*

- Right click on the *Geometry* option and select the *Rectangle*.
- In the *Rectangle* environment, select *Size* with *Width: 0.5 and Height: 2*. Click on the *Build All* option at the upper part of tool bar. A rectangle graph will appear on the right side of the application window.

*Step 3*

- In the *Laminar Flow* node, select the *Compressibility:Incompressible Flow*.

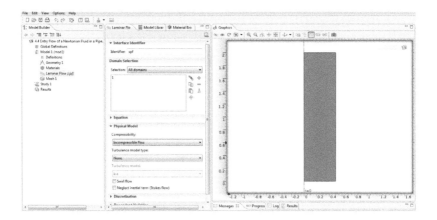

- Now in the *Model Builder*, select *Fluid Properties 1* option available in *Laminar Flow*. This will open a tab to enter the coefficients of the characteristics of a model equation.
- In the *Domain Selection* panel, you will see an equation of the form

$$\rho\left(u.\nabla\right)u = \nabla \cdot \left[-pl + \mu\left(\nabla u + (\nabla u)^{T}\right)\right] + F \tag{4.16}$$

$$\rho\nabla \cdot u = 0. \tag{4.17}$$

To convert Equations (4.16) and (4.17) to the desired form of Equations (4.9) and (4.10), respectively, the value of coefficients in Equations (4.16) and (4.17) to be changed as follows:

$\rho = 10$, $\mu = $ eta.

Then, click on the "+" sign at the top right corner, this will add *Domain 1* in the *Domain Selection* tab.

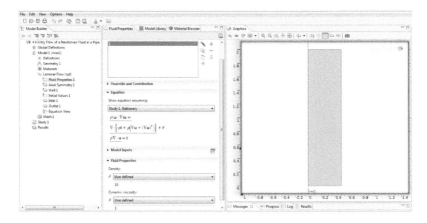

- Select *Parameters* by right clicking on the *Global Definitions* tab. Another window will open namely *Parameters*. Click on it and define *etainf: 5×10⁻²*, *eta0: 492×10⁻³*, *lambda: 1 × 10⁻¹*, *n:4×10⁻¹*.

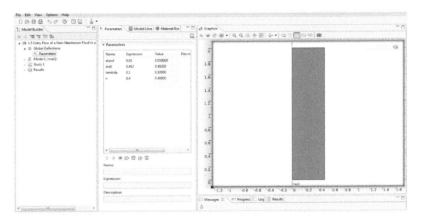

- Select *Variables* by right clicking on the *Definitions* tab. Another window will open namely *Variables 1*. Click on it and define:

$$eta = etainf + (eta0 - etainf) \times (1 + (lambda \times gammadot)^2)^{(n-1)/2}$$

$$gammadot2 = 2 \times \left( ur^2 + \left( \frac{u}{r} \right)^2 + wz^2 \right) + (uz + wr)^2$$

$$gammadot = sqrt(gammadot2)$$

Also, the COMSOL automatically calculates "*gammadot*" if the variable used is "*mod1.spf.sr.*"

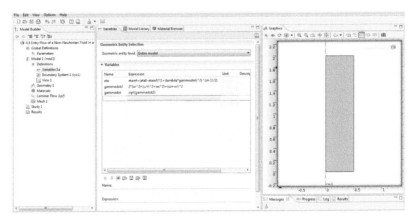

*Step 4*

- Right click on the *Laminar Flow* tab in *Model Builder.* Select the *Inlet* option. Another window will open namely *Inlet 1.* Click on it.
- Select the bottom boundary on the rectangle graph. Click on the "+" sign at the top right corner, this will add *boundary 2* in the *Boundary Selection* tab. At this point, put *Normal inflow velocity* $U_0$: 1. This step will add a boundary condition as given in Equation (4.11).

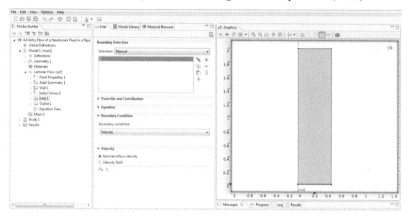

- Again right click on the *Laminar Flow* tab in *Model Builder.* Select the *Outlet* option. Another window will open namely *Outlet 1.* Click on it.
- Select the top boundary on the rectangle graph. Click on the "+" sign at the top right corner, this will add *boundary 3* in the *Boundary Selection* tab. At this point, put *Pressure* $p_0$: 0. This step will add a boundary condition as given in Equation (4.12).

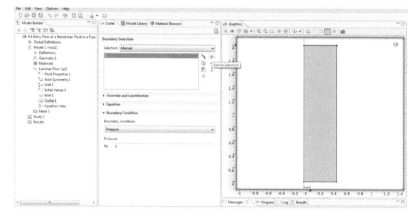

- Now, in the *Model Builder*, select *Wall 1* option available in *Laminar Flow*. By default it has *boundary 4* in the *Boundary Selection* tab. At this point, put boundary condition to *No slip*. This step will add a boundary condition as given in Equation (4.14).

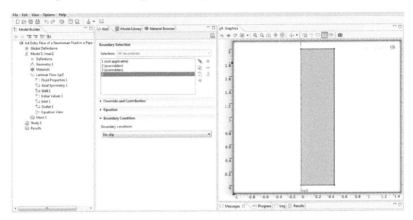

- Now, in the *Model Builder*, select *Axial Symmetry 1* option available in *Laminar Flow*. By default, it has *boundary 1* in the *Boundary Selection* tab. In COMSOL, by default, the axisymmetric geometry allows slip and this step will add a boundary condition as given in Equation (4.13).

*Step 5*

- Click on the *Mesh* option in *Model Builder*. Select *Normal Mesh* Type. Click on *Build All* option at the top of ribbon. A dialogue box will appear in the *Message* tab as: "*Complete mesh consists of 3353 elements.*"

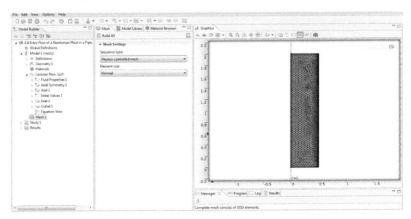

- Now, go to *Study 1* option in the model pellet tab. Click on the *Compute* *(=)* button. A plot of velocity magnitude, $U = \text{sqrt}(u2 + w2)$ will appear.
- Save the simulation.

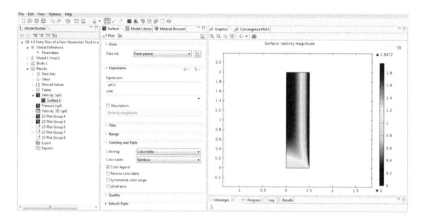

## Step 6

- To plot the inlet velocity, click on the *Line Graph 1* tab. Another *Plot* window will open. Click on the "+" sign in the *y*-axis data and select *Velocity magnitude, U.*
- Select the bottom boundary on the rectangle graph. Click on the "+" sign at the top right corner, this will add *boundary 2* in the *Boundary Selection* tab. Click on the "+" sign in the *x*-axis data and select *r-coordinate (r)*. Plot the *Line Graph*. A graph will appear giving the profile of *Velocity magnitude versus r* for boundary 2.

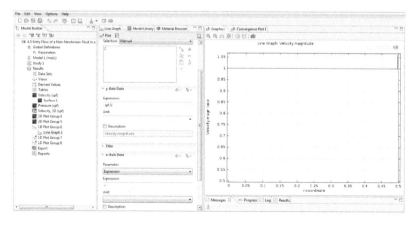

- Now, select the top boundary on the rectangle graph. Click on the "+" sign at the top right corner, this will add *boundary 3* in the *Boundary Selection* tab. Click on the "+" sign in the *x*-axis data and select *r-coordinate (r)*. Plot the *Line Graph*. A graph will appear giving the profile of *Velocity magnitude versus r* for boundary 3 as shown in Figure 4.4.

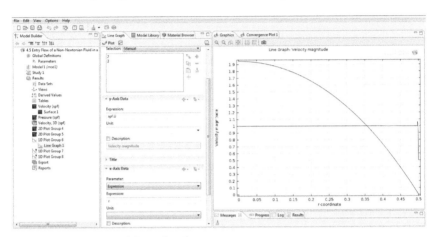

**FIGURE 4.4**    Velocity at inlet and outlet.

- Now select the left boundary on the rectangle graph. Click on the "+" sign at the top right corner, this will add *boundary 1* in the *Boundary Selection* tab. Click on the "+" sign in the *x*-axis data and select *z-coordinate (z)*. Plot the *Line Graph*. A graph will appear giving the profile of *Velocity magnitude versus z* for boundary 1 as shown in Figure 4.5.

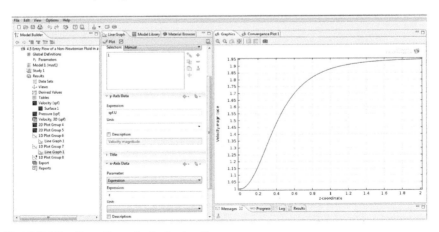

**FIGURE 4.5**    Centerline velocity in pipe flow.

## 4.4 SIMULATION OF FLOW IN MICROFLUIDIC DEVICES

### *4.4.1 PROBLEM STATEMENT*

Consider the flow through *T*-sensor generally applied in microfluidic medical devices as shown in Figure 4.6. It is represented in COMSOL by two rectangles with the following dimensions:

Rectangle 1: Width = 2; Height = 0.5;
and Base corner: at $x = -0.5$, $y = -0.25$.

Rectangle 2: Width = 0.5; Height = 1.5;
and Base corner: at $x = -1$, $y = -0.75$.

It is described mathematically by Navier–Stokes equation (4.18) and continuity equation (4.19) in the dimensionless form. An order of magnitude analysis reveals (Schlichting, 1979) that *z*-momentum in 2D boundary layer flow can be neglected. Therefore, in Cartesian coordinates, 2D, laminar, incompressible flow with constant viscosity is described by

$$\rho\left( u\frac{\partial u}{\partial x} + v\frac{\partial u}{\partial y} \right) = -\frac{\partial p}{\partial x} + \mu\left( \frac{\partial^2 u}{\partial x^2} + \frac{\partial^2 u}{\partial y^2} \right) \tag{4.18}$$

$$\frac{\partial u}{\partial x} + \frac{\partial v}{\partial y} = 0 \tag{4.19}$$

Boundary conditions:
Inflow from the top:

At y = 1.5, $u = v =$ at any x. $\tag{4.20}$

Inflow from the bottom:

At y = 0, $u = v = 1$ at any x. $\tag{4.21}$

At the outflow, the viscous stress in fully developed flow is considered; hence, pressure is zero.

At $x = 2.5$, $P_0 = 0$ at any y. $\tag{4.22}$

The wall at all other boundaries is having no slip condition with the velocity zero for both components.
Perform the overall mass balance.
Given $\rho = 1$, $\mu = 1$.

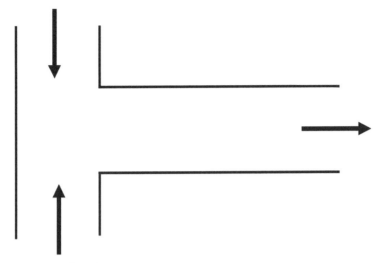

**FIGURE 4.6** Flow in a *T*-sensor.

### 4.4.2 SIMULATION APPROACH

*Step 1*

- Open COMSOL Multiphysics.
- Select *2D Space Dimension* from the list of options. Hit the *next* arrow at the upper right corner.

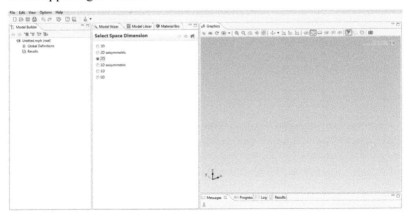

- Select and expand the *Fluid Flow* folder from the list of options in *Model Wizard* and select the *Laminar Flow*. Again hit the *next* arrow at the upper right corner.

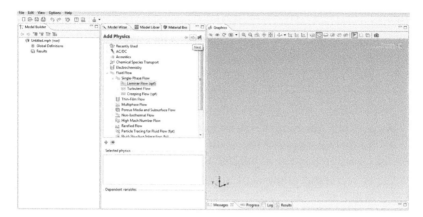

- Then, select *Stationary* from the list of *Study Type* options, and click on the *Finish flag* at the upper right corner of the application.

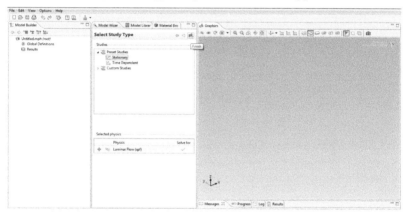

- Select the *Root* node and change the *Unit System: None*.

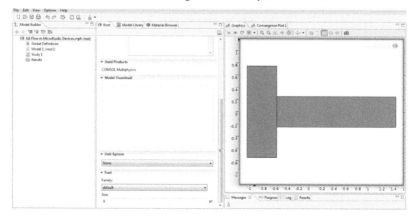

*Step 2*

- Right click on the *Geometry* option and select the *Rectangle*.
- In the *Rectangle 1* environment, select the dimension as mentioned above. Click on the *Build All* option at the upper part of tool bar. A rectangle graph will appear on the right side of the application window.
- Again right click on the *Geometry* option and select the *Rectangle*.
- In the *Rectangle 2* environment, select the dimension as mentioned above. Click on the *Build All* option at the upper part of tool bar. A rectangle graph will appear on the right side of the application window.

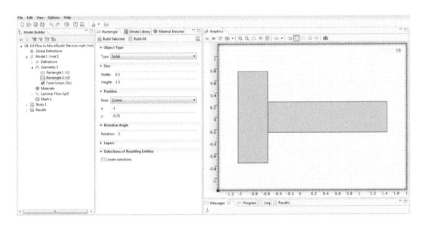

*Step 3*

- Now, in the *Model Builder*, select *Fluid Properties 1* option available in *Laminar Flow*. This will open a tab to enter the coefficients of the characteristics of a model equation.
- In the *Domain Selection* panel, you will see an equation of the form

$$\rho(u.\nabla)u = \nabla \cdot \left[ -pl + \mu\left( \nabla u + (\nabla u)^T - \frac{2}{3}\mu(\nabla \cdot u)l \right) \right] + F \qquad (4.23)$$

$$\nabla \cdot (\rho u) = 0. \qquad (4.24)$$

To convert Equations (4.23) and (4.24) to the desired form of Equations (4.18) and (4.19), respectively, the value of coefficients in Equations (4.18) and (4.19) to be changed as follows:

$\rho = 10, \mu = 1$.

Then, click on the "+" sign at the top right corner; this will add *Domain 1* in the *Domain Selection* tab.

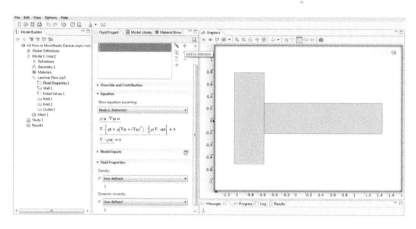

*Step 4*

- Now, right click on the *Laminar Flow* tab in *Model Builder*. Select the *Inlet* option. Another window will open namely *Inlet 1*. Click on it.
- Select the top boundary on the Rectangle 2 graph. Click on the "+" sign at the top right corner, this will add *boundary 3* in the *Boundary Selection* tab. At this point, put *Normal inflow velocity* $U_0$: *1*. This step will add a boundary condition as given in Equation (4.20).

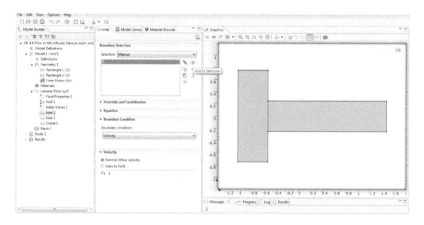

- Again right click on the *Laminar Flow* tab in *Model Builder*. Select the *Inlet* option. Another window will open namely *Inlet 2*. Click on it.

- Select the bottom boundary on the Rectangle 2 graph. Click on the "+" sign at the top right corner, this will add *boundary 2* in the *Boundary Selection* tab. At this point, put *Normal inflow velocity $U_0$: 1*. This step will add a boundary condition as given in Equation (4.21).

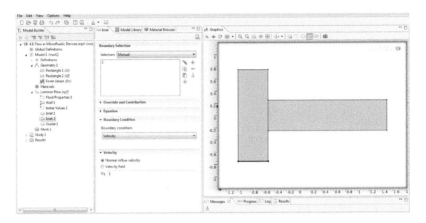

- Again right click on the *Laminar Flow* tab in *Model Builder*. Select the *Outlet* option. Another window will open namely *Outlet 1*. Click on it.
- Select the right boundary on the Rectangle 1 graph. Click on the "+" sign at the top right corner; this will add *boundary 9* in the *Boundary Selection* tab. At this point, put *Pressure $p_0$: 0*. This step will add a boundary condition as given in Equation (4.22).

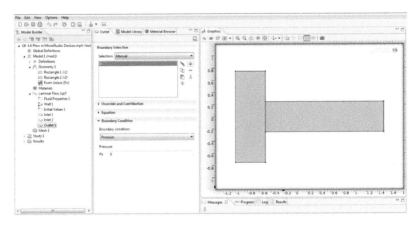

- Now, in the *Model Builder*, select *Wall 1* option available in *Laminar Flow*. By default, it has all other boundaries in the *Boundary Selection* tab. At this point, put boundary condition to *No slip*.

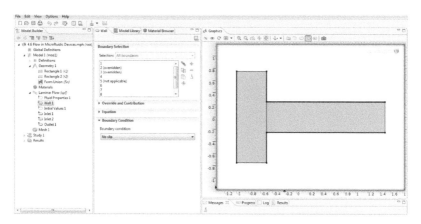

*Step 5*

- Click on the *Mesh* option in *Model Builder*. Select *Normal Mesh* Type. Click on *Build All* option at the top of ribbon. A dialogue box will appear in the *Message* tab as: *"Complete mesh consists of 2101 elements"* as shown in Figure 4.7.

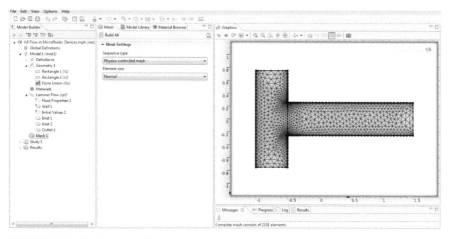

**FIGURE 4.7** Meshing in a *T*-sensor.

- Now, go to *Study 1* option in the model pellet tab. Click on the *Compute (=)* button.
- Save the simulation.

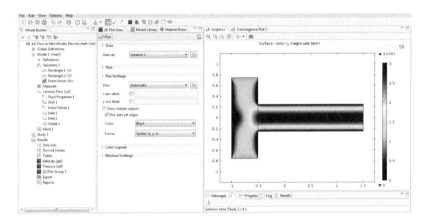

*Step 6*

- Right click on the *Results* tab and choose the *2D Plot Group* option. Another window will open namely *2D Plot Group 1.* Select *Streamline* by right clicking on it. Another window will open namely *Streamline 1* as shown in Figure 4.8.

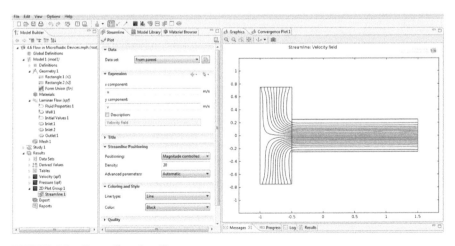

**FIGURE 4.8**   Streamlines in a *T*-sensor.

*Step 7*

- To calculate the overall mass balance, select *Derived Values* option available in the *Results* tab.
- Select the *Line Integration* option by right clicking on *Derived Values* tab.
- Select the top inlet boundary on the Rectangle 2 graph. Click on the "+" sign at the top right corner, this will add *boundary 3* in the *Boundary Selection* tab. Click on "= Evaluate" button at the top. The velocity magnitude is 0.49575.

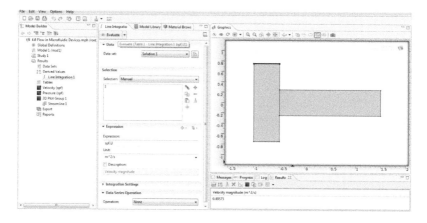

- Now, select the bottom inlet boundary on the Rectangle 2 graph. Click on the "+" sign at the top right corner, this will add *boundary 2* in the *Boundary Selection* tab. Click on "= Evaluate" button at the top. The velocity magnitude is 0.49563.

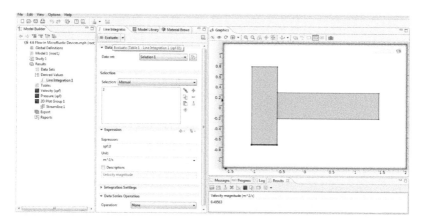

- Now, select the right outlet boundary on the Rectangle 1 graph. Click on the "+" sign at the top right corner, this will add *boundary 9* in the *Boundary Selection* tab. Click on "= Evaluate" button at the top. The velocity magnitude is 0.97209.
- The sum of inlet is 0.99138 (0.49575 + 0.49563 = 0.99138) and 0.97209 out. This is accurate to within 1.8%.

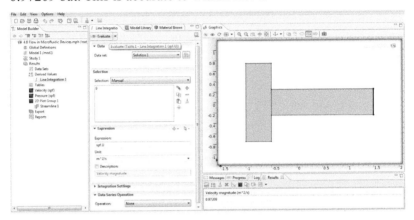

- To further improve the solution, change the mesh element size from *Normal Mesh* type to *Extremely fine Mesh* type. Click on *Build All* option at the top of ribbon. A dialogue box will appear in the *Message* tab as: "*Complete mesh consists of 27,444 elements.*"

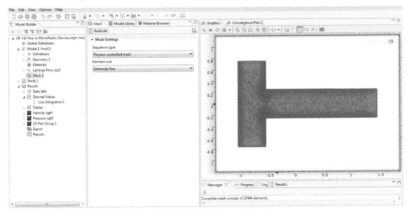

- Again calculate the inlet and outlet flow rates as done before. The sum of inlet flow rates is 0.99712 (0.49858 + 0.49854 = 0.99712) and 0.99713 out. The error here is 0.0%. This solution is obtained using *Extremely fine Mesh* having 27,444 elements, while the earlier solution was obtained using *Normal Mesh* having 2101 elements.

## 4.5 SIMULATION OF TURBULENT FLOW IN A PIPE

### 4.5.1 *PROBLEM STATEMENT*

The entry flow of a Newtonian fluid into a pipe, as shown in Figure 4.1, is described mathematically by Navier–Stokes equation (4.25) and continuity equation (4.26). An order of magnitude analysis reveals (Schlichting, 1979) that *z*-momentum in 2D boundary layer flow can be neglected. Therefore, in Cartesian coordinates, 2D, turbulent, incompressible flow with constant viscosity is described by

$$\rho\left(u\frac{\partial u}{\partial r} + w\frac{\partial u}{\partial z}\right) = -\frac{\partial p}{\partial r} + \mu\left(\frac{\partial^2 u}{\partial r^2} + \frac{\partial^2 u}{\partial z^2}\right) \tag{4.25}$$

$$\frac{\partial u}{\partial r} + \frac{\partial w}{\partial z} = 0. \tag{4.26}$$

Boundary conditions:

At $z = 0$, $u = 1$; $w = 1$ at any $r$. $\tag{4.27}$
At $z = 0.1$, $p_0 = 0$ at any $r$. $\tag{4.28}$

By default, the centerline has slip condition with velocity zero for $r$ component and nonzero or floating for $z$ component; hence, we have the following:

At $r = 0$, $u = 0$, $w$ is nonzero or floating at any z. $\tag{4.29}$

For the wall at other boundary is having sliding wall condition with the velocity zero for both components, hence

At $r = 0.025$, $u = 0$; $w = 0$ at any $z$. $\tag{4.30}$
Given $\rho = 110^3$ kg/m$^3$, $\mu = 1 \times 10^{-3}$ Pa s.

### 4.5.2 *SIMULATION APPROACH*

*Step 1*

- Open COMSOL Multiphysics.
- Select *2D axisymmetric Space Dimension* from the list of options. Hit the *next* arrow at the upper right corner.

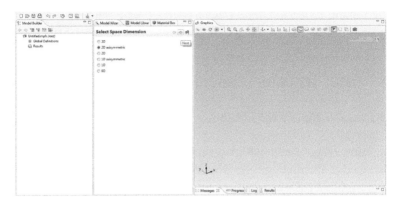

- Select and expand the *Fluid Flow* folder from the list of options in *Model Wizard* and click on the *Single Phase Flow* and select the *Turbulent Flow, k–ε*. Again hit the *next* arrow.

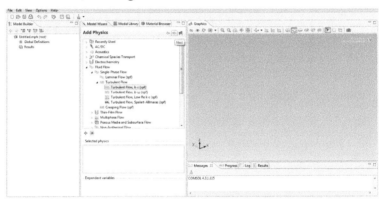

- Then, select *Stationary* from the list of *Study Type* options, and click on the *Finish flag* at the upper right corner of the application.

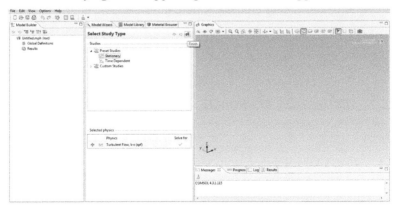

*Step 2*

- Right click on the *Geometry* option and select the *Rectangle*.
- In the *Rectangle* environment, select *Size* with *Width: 0.025 and Height: 0.1*. Click on the *Build All* option at the upper part of tool bar. A rectangle graph will appear on the right side of the application window.

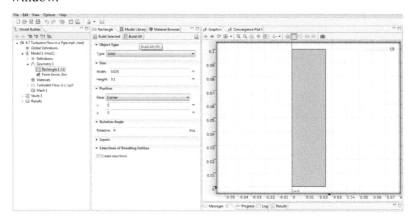

*Step 3*

- In the *Laminar Flow* node, select the *Compressibility:Incompressible Flow*.

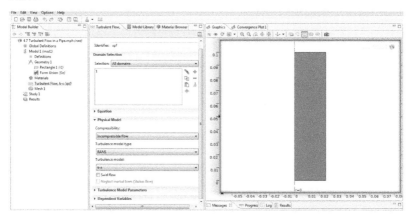

- Now, in the *Model Builder*, select *Fluid Properties 1* option available in *Turbulent Flow*. This will open a tab to enter the coefficients of the characteristics of a model equation.

- In the *Domain Selection* panel, you will see an equation of the form

$$\rho\left(u\cdot\nabla\right)u = \nabla\cdot\left[-pl+\left(\mu+\mu_T\right)\left(\nabla u+\left(\nabla u\right)^T-\frac{2}{3}\rho kl\right)\right]+F \quad (4.31)$$

$$\rho\,\nabla\cdot u = 0 \quad (4.32)$$

$$\rho\left(u\cdot\nabla\right)k = \nabla\cdot\left[\left(\mu+\frac{\mu_T}{\sigma_k}\right)\nabla k\right]+P_k-\rho\in \quad (4.33)$$

$$\rho\left(u\cdot\nabla\right)\in = \nabla\cdot\left[\left(\mu+\frac{\mu_T}{\sigma_\in}\right)\nabla\in\right]+Cc_1\frac{\in}{k}P_k-Cc_2\rho\frac{\in^2}{k},\in = ep. \quad (4.34)$$

To convert Equations (4.25) and (4.26) to the desired form of Equations (4.31) and (4.32), respectively, the value of coefficients in Equations (4.25) and (4.26) to be changed as follows:

$\rho = 1\times10^3$ kg/m³, $\mu = 1\times10^{-3}$ Pa s.

Then, click on the "+" sign at the top right corner; this will add *Domain 1* in the *Domain Selection* tab.

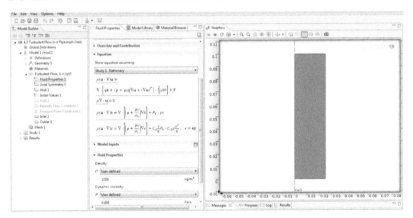

*Step 4*

- Now, right click on the *Turbulent Flow* tab in *Model Builder*. Select the *Inlet* option. Another window will open namely *Inlet 1*. Click on it.
- Select the bottom boundary on the rectangle graph. Click on the "+" sign at the top right corner; this will add *boundary 2* in the *Boundary Selection* tab. At this point, put *Normal inflow velocity $U_0$: 1*. This step will add a boundary condition as given in Equation (4.27).

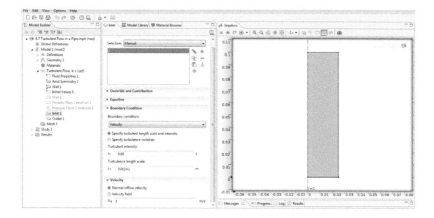

- Again right click on the *Turbulent Flow* tab in *Model Builder*. Select the *Outlet* option. Another window will open namely *Outlet 1*. Click on it.
- Select the top boundary on the rectangle graph. Click on the "+" sign at the top right corner; this will add *boundary 3* in the *Boundary Selection* tab. At this point, put *Pressure $p_0$: 0*. This step will add a boundary condition as given in Equation (4.28).

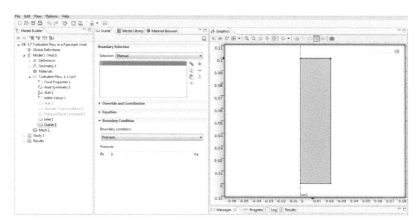

- Now, in the *Model Builder*, select *Wall 1* option available in *Laminar Flow*. By default, it has *boundary 4* in the *Boundary Selection* tab. At this point, put boundary condition to *Sliding Wall*. This step will add a boundary condition as given in Equation (4.30).

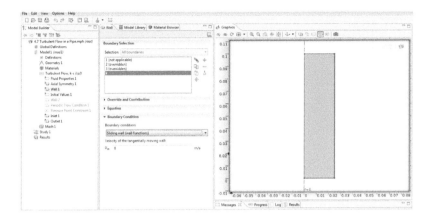

- Now, in the *Model Builder*, select *Axial Symmetry 1* option available in *Turbulent Flow*. By default, it has *boundary 1* in the *Boundary Selection* tab. In COMSOL, by default, the axisymmetric geometry allows slip, and this step will add a boundary condition as given in Equation (4.29).

*Step 5*

- Click on the *Mesh* option in *Model Builder*. Select *Normal Mesh* Type. Click on *Build All* option at the top of ribbon. A dialogue box will appear in the *Message* tab as: "*Complete mesh consists of 6931 elements.*"

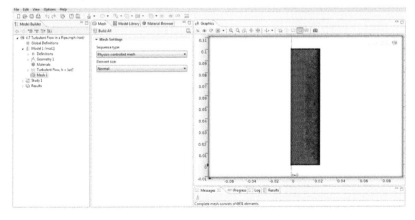

- Now, go to *Study 1* option in the model pellet tab. Click on the *Compute (=)* button. A plot of velocity magnitude, U will appear.
- Save the simulation.

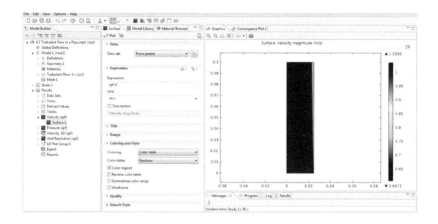

**Step 6**

- To plot the outlet velocity, select the *Line Graph 1* tab. Another *Plot* window will open. Click on the "+" sign in the *y*-axis data and select *Velocity magnitude*, U.
- Select the bottom boundary on the rectangle graph. Click on the "+" sign at the top right corner, this will add *boundary 3* in the *Boundary Selection* tab. Click on the "+" sign in the *x*-axis data and select *r-coordinate (r)*. Plot the *Line Graph*. A graph will appear giving the profile of *Velocity magnitude versus r* for boundary 3 as shown in Figure 4.9.

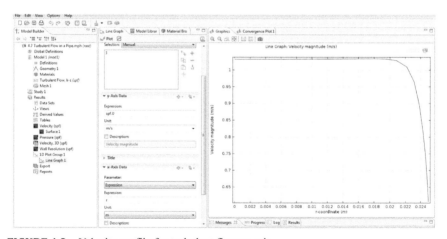

**FIGURE 4.9** Velocity profile for turbulent flow at exit.

- To plot the turbulent kinetic energy (*k*) as a function of radius, change the *Expression* in the *y-axis Data* from spf.U to *k*. Now, go to *Study* option in the model pellet tab. Click on the *Compute (=)* button. A graph will appear giving the profile of *Viscosity versus Radius*.

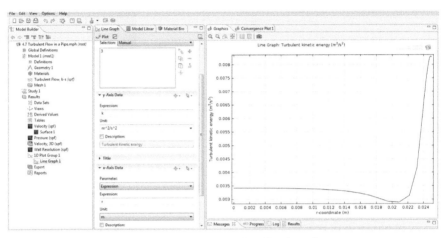

**FIGURE 4.10**   Turbulent kinetic energy for turbulent flow in a pipe.

- To plot the turbulent dissipation rate ($\in$ or ep) as a function of radius, change the *Expression* in the *y-axis Data* from $k$ to ep. Now, go to *Study* option in the model pellet tab. Click on the *Compute* (=) button. A graph will appear giving the profile of *Viscosity versus Radius* as shown in Figure 4.11.

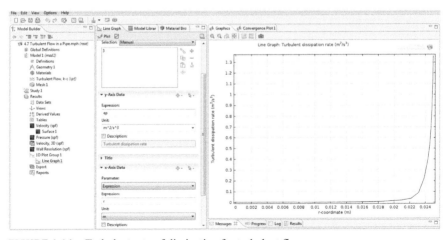

**FIGURE 4.11**   Turbulent rate of dissipation for turbulent flow.

- To plot the Pressure (*p*), again select the *Line Graph 1* tab. Click on the "+" sign in the *y*-axis data and select *Pressure (p)*.
- Select the left boundary on the rectangle graph. Click on the "+" sign at the top right corner; this will add *boundary 1* in the *Boundary Selection* tab. Change the *x*-axis data to Arc length. Click on *Plot* icon in the *Line Graph*. A graph will appear giving the profile of *Pressure (p) versus Arc length* for centerline as shown in Figure 4.12.

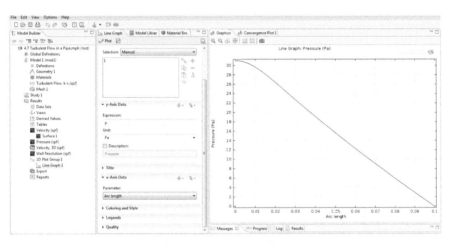

**FIGURE 4.12** Pressure in turbulent flow along centerline.

## 4.6 SIMULATION OF START-UP FLOW IN A PIPE

The entry flow of a Newtonian fluid into a pipe as shown in Figure 4.1 is described mathematically by Navier–Stokes equation (4.35) and continuity equation (4.36).

It is represented in COMSOL by rectangle with the following dimensions: Rectangle 1: Width = 0.0025, Height = 0.025, Base corner: at $r = 0, z = 0$.

An order of magnitude analysis reveals (Schlichting, 1979) that $z$-momentum in 2D boundary layer flow can be neglected. Therefore, in Cartesian coordinates, 2D, laminar, incompressible flow with constant viscosity is described by

$$\rho \frac{\partial u}{\partial t} + \rho \left( u \frac{\partial u}{\partial r} + w \frac{\partial u}{\partial z} \right) = -\frac{\partial p}{\partial r} + \mu \left( \frac{\partial^2 u}{\partial r^2} + \frac{\partial^2 u}{\partial z^2} \right) \qquad (4.35)$$

$$\frac{\partial u}{\partial r} + \frac{\partial w}{\partial z} = 0. \tag{4.36}$$

Boundary conditions:

$$\Delta p = 2.5 \; Pa \; at \; any \; r \tag{4.37}$$

where, $\Delta p$ is the pressure difference between inlet and outlet boundary

By default, the centerline boundary condition is in axisymmetric geometry having slip condition, with the velocity being zero for $r$ component and nonzero or floating for $z$ component; hence, we have the following:

At $r = 0$, $u = 0$, $w$ is nonzero or floating at any $z$. $\tag{4.38}$

For the wall at other boundary is having no slip condition with the velocity zero for both components, hence

At $r = 0.0025$, $u = w\ 0$ at any $z$. $\tag{4.39}$

Perform the simulation till 7 s and plot the velocity as a function of radius and time. Given: $\rho = 1\times10^3$ kg/m³, $\mu = 1\times10^{-3}$ Pa s.

### 4.6.1  SIMULATION APPROACH

*Step 1*

- Open COMSOL Multiphysics.
- Select *2D* axisymmetric *Space Dimension* from the list of options. Hit the *next* arrow at the upper right corner.

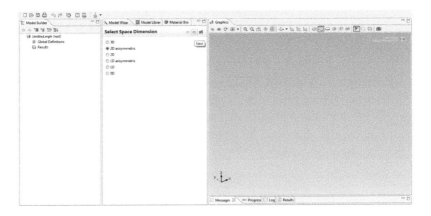

- Select and expand the *Fluid Flow* folder from the list of options in *Model Wizard* and select the *Laminar Flow*. Again hit the *next* arrow at the upper right corner.

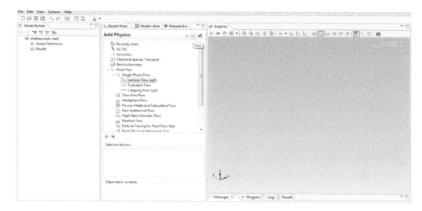

- Then, select *Time Dependent* from the list of *Study Type* options, and click on the *Finish flag* at the upper right corner of the application.

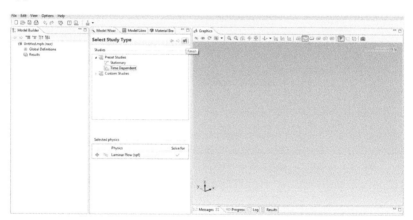

*Step 2*

- Right click on the *Geometry* option and select the *Rectangle*.
- In the *Rectangle* environment, select *Size* with *Width: 0.0025 and Height: 0.025*. Click on the *Build All* option at the upper part of tool bar. A rectangle graph will appear on the right side of the application window.

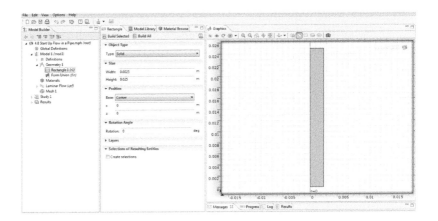

## Step 3

- In the *Laminar Flow* node, select the *Compressibility:Incompressible Flow* and Discretization to "P2+P1".

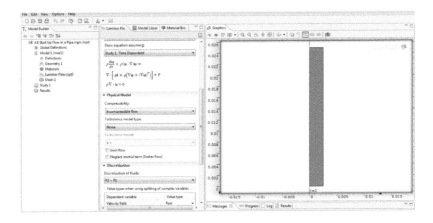

- Now, in the *Model Builder*, select *Fluid Properties 1* option available in *Laminar Flow*. This will open a tab to enter the coefficients of the characteristics of a model equation.
- In the *Domain Selection* panel, you will see an equation of the form

$$\rho\frac{\partial u}{\partial t} + \rho(u \cdot \nabla)u = \nabla \cdot \left[ -pl + \mu\left(\nabla u + (\nabla u)^T\right) \right] + F \qquad (4.40)$$

$$\rho\nabla \cdot u = 0 \qquad (4.41)$$

To convert Equations (4.40) and (4.41) to the desired form of Equations (4.35) and (4.36), respectively, the value of coefficients in Equations (4.35) and (4.36) to be changed as follows:

$\rho = 1 \times 10^3$ kg/m$^3$, $\mu = 1 \times 10^{-3}$ Pa s.

Then, click on the "+" sign at the top right corner, this will add *Domain 1* in the *Domain Selection* tab.

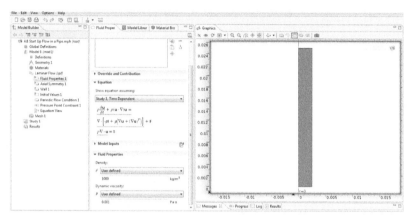

## Step 4

- Now, right click on the *Laminar Flow* tab in *Model Builder*. Select the *Periodic Flow Condition* option. Another window will open namely *Periodic Flow Condition 1*. Click on it.
- Select the bottom and top boundary on the rectangle graph. Click on the "+" sign at the top right corner, this will add *boundaries 2 and 3* in the *Boundary Selection* tab. At this point, put *Pressure Difference* $p_{src} - p_{dst}$: 2.5. This step will add a boundary condition as given in Equation (4.37).

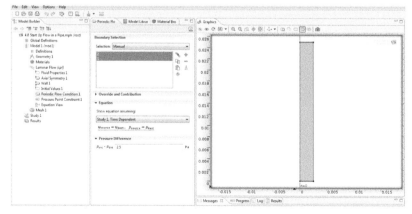

- Now, in the *Model Builder*, select *Wall 1* option available in *Laminar Flow*. By default, it has *boundary 4* in the *Boundary Selection* tab. At this point, put boundary condition to *No slip.* This step will add a boundary condition as given in Equation (4.39).

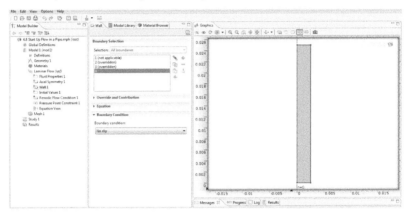

- Now, in the *Model Builder*, select *Axial Symmetry 1* option available in *Laminar Flow*. By default, it has *boundary 1* in the *Boundary Selection* tab. In COMSOL, by default, the axisymmetric geometry allows slip, and this step will add a boundary condition as given in Equation (4.38).
- Now, right click on the *Laminar Flow* tab in *Model Builder.* Select the *Pressure Point Constraint* option. Another window will open namely *Pressure Point Constraint 1.* Click on it.
- Select the point 4 at the outlet on the rectangle graph. Click on the "+" sign at the top right corner; this will add *point 4* in the *Point Selection* tab. At this point, put *Pressure $p_0$: 0.*

*Step 5*

- Click on the *Mesh* option in *Model Builder.* Select *Normal Mesh* Type. Click on *Build All* option at the top of ribbon. A dialogue box will appear in the *Message* tab as: *"Complete mesh consists of 8453 elements."*

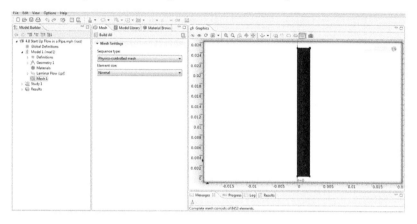

- Now, go to *Study 1, Step1: Time Dependent* option in the model pellet tab and set the *Times: 0 0.1 0.2 0.3 0.4 0.5 0.6 0.7 0.8 0.9 1.0 2 3 4 5 6 7*. Click on the *Compute (=)* button. A plot of velocity magnitude *U* will appear.
- Save the simulation.

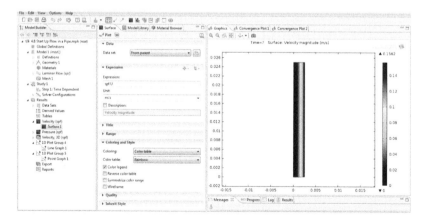

*Step 6*

- To plot the inlet velocity, select the *Line Graph 1* tab. Another *Plot* window will open. Click on the "+" sign in the *y*-axis data and select *Velocity magnitude, U.*

- Select the bottom boundary on the rectangle graph. Click on the "+" sign at the top right corner, this will add *boundary 2* in the *Boundary Selection* tab. Click on the "+" sign in the *x*-axis data and select *r-coordinate (r)*. Plot the *Line Graph*. A graph will appear giving the profile of *Velocity magnitude versus r* for boundary 2 as shown in Figure 4.13.

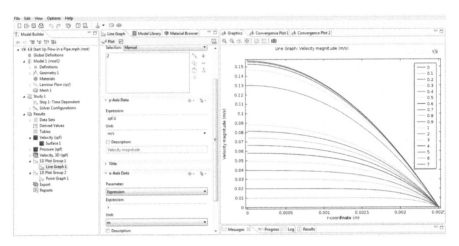

**FIGURE 4.13**    Transient velocity profiles.

- Now, select the top boundary on the rectangle graph. Click on the "+" sign at the top right corner, this will add *boundary 3* in the *Boundary Selection* tab. Click on the "+" sign in the *x*-axis data and select *r-coordinate (r)*. Plot the *Line Graph*. The same graph as shown in Figure 4.13 will appear giving the profile of *Velocity magnitude versus r* for boundary 3.
- To plot the velocity as function of time, select the *Point Graph 1* in the *1D Plot Group* tab. Another *Plot* window will open. Click on the "+" sign in the *y*-axis data and select *Velocity magnitude, U*.
- Choose the point at *r* = 0, *z* = 0 on the rectangle graph. Click on the "+" sign at the top right corner, this will add *point 1* in the *Selection* tab. Set the *x*-axis data as *Parameter:Time*. Click on *Plot* icon in the *Point Graph*. A graph will appear giving the profile of *Velocity magnitude versus Time* for point 1. Figure 4.14 shows that steady state is reached within 7 s.

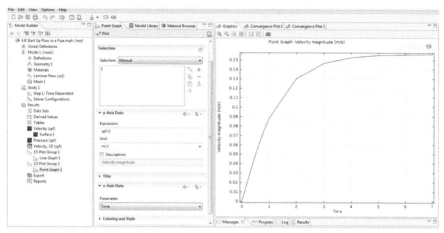

**FIGURE 4.14**  Velocity at $r = 0$ and $z = 0$.

## 4.7  SIMULATION OF FLOW THROUGH AN ORIFICE

The flow through the orifice is represented in COMSOL by two rectangles with the following dimensions:

Rectangle 1: Width = 3; Height = 16; and Base corner: at $r = 0$, $z = 0$.
Rectangle 2: Width = 2.5; Height = 0.15; and Base corner: at $r = 0$, $z = 3.925$.

Make the diameter of the orifice hole equal to 1, and thickness of the orifice plate/diameter of the hole is 0.15. It is described mathematically by Navier–Stokes equation (4.42) and continuity equation (4.43) in the dimensionless form. An order of magnitude analysis reveals (Schlichting, 1979) that $z$-momentum in the 2D boundary layer flow can be neglected. Therefore, in Cartesian coordinates, 2D, laminar, incompressible flow with constant viscosity is described by

$$\rho\left(u\frac{\partial u}{\partial r} + w\frac{\partial u}{\partial z}\right) = -\frac{\partial p}{\partial r} + \mu\left(\frac{\partial^2 u}{\partial r^2} + \frac{\partial^2 u}{\partial z^2}\right) \tag{4.42}$$

$$\frac{\partial u}{\partial r} + \frac{\partial w}{\partial z} = 0. \tag{4.43}$$

Boundary conditions:

$$\text{At } z = 0, u = \left(\frac{2}{36}\right) \times \left(1 - \left(\frac{r}{3}\right)^2\right), \; w = 0 \text{ at any } r \tag{4.44}$$

$$\text{At } z = 16, \, p_0 = 0 \text{ at any } r. \tag{4.45}$$

By default, the centerline has slip condition with velocity zero for $r$ component and nonzero or floating for $z$ component; hence, we have the following:

$$\text{At } r = 0, \, u = 0, \, w \text{ is nonzero or floating at any } z. \tag{4.46}$$

For the wall at other boundary is having no slip condition with the velocity zero for both components, hence

$$\text{At } r = 3, \, u = w = 0 \text{ at any } z. \tag{4.47}$$

Given: $\rho = 1$, $\mu = 1/10^x$.

Perform the parametric study by varying the Reynolds number from 1 to about 30.

### 4.7.1 SIMULATION APPROACH

*Step 1*

- Open COMSOL Multiphysics.
- Select *2D* axisymmetric *Space Dimension* from the list of options. Hit the *next* arrow at the upper right corner.

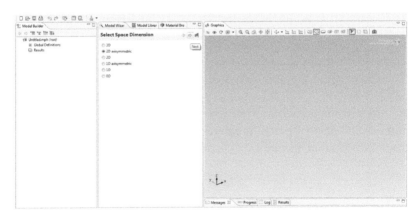

- Select and expand the *Fluid Flow* folder from the list of options in *Model Wizard* and select the *Laminar Flow*. Again hit the *next* arrow at the upper right corner.

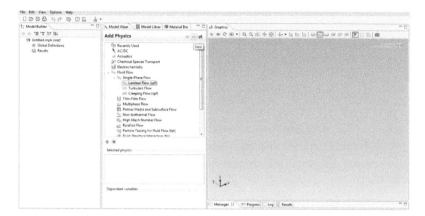

- Then, select *Stationary* from the list of *Study Type* options, and click on the *Finish flag* at the upper right corner of the application.

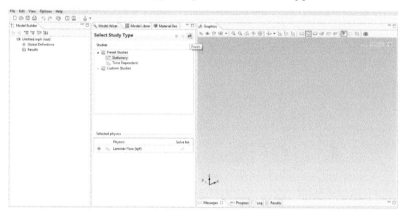

- Select the *Root* node and change the *Unit System: None*.

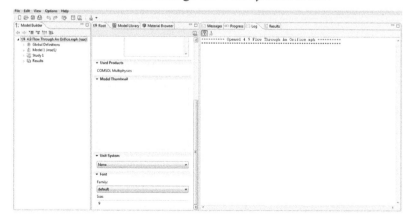

*Step 2*

- Right click on the *Geometry* option and select the *Rectangle*.
- In the *Rectangle 1* environment, select dimensions as mentioned above. Click on the *Build All* option at the upper part of tool bar. A rectangle graph will appear on the right side of the application window.
- Again right click on the *Geometry* option and select the *Rectangle*.
- In the *Rectangle 2* environment, select dimensions as mentioned above. Click on the *Build All* option at the upper part of tool bar. A rectangle graph will appear on the right side of the application window.

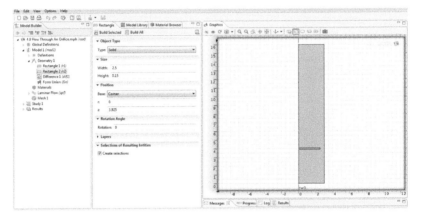

- Right click on the *Geometry* option and select the *Boolean Operations* taskbar and click *Difference*.
- In the *Difference* environment, select *Objects to add* as "$r_1$" and *Objects to subtract* as "$r_2$." Check "Keep input objects" and click on the *Build All* option at the upper part of the tool bar.

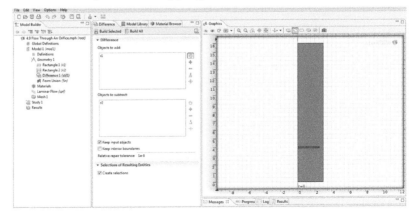

*Step 3*

- In the *Laminar Flow* node, select the *Compressibility: Incompressible Flow* and the Discretization to P2+P1.

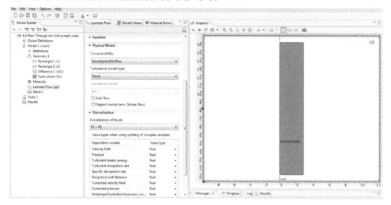

- Now, in the *Model Builder*, select *Fluid Properties 1* option available in *Laminar Flow*. This will open a tab to enter the coefficients of the characteristics of a model equation.
- In the *Domain Selection* panel, you will see an equation of the form

$$\rho(u \cdot \nabla)u = \nabla \cdot \left[ -\rho l + \mu \left( \nabla u + (\nabla u)^T \right) \right] + F \tag{4.48}$$

$$\rho \nabla \cdot u = 0 \tag{4.49}$$

To convert Equations (4.48) and (4.49) to the desired form of Equations (4.42) and (4.43), respectively, the value of coefficients in Equations (4.48) and (4.49) to be changed as follows:
$\rho = 10$, $\mu = 1/\text{Re}$.
Then, click on the "+" sign at the top right corner; this will add *Domain 1* in the *Domain Selection* tab.

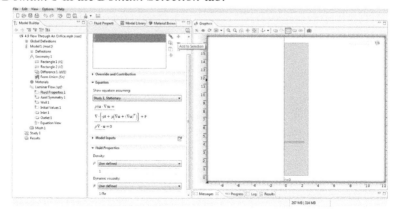

- Select *Variables* by right clicking on the *Definitions* tab. Another window will open namely *Variables*. Click on it and define *Re: 10ˣ*.

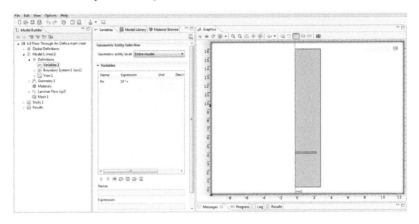

- Select *Parameters* by right clicking on the *Global Definitions* tab. Another window will open namely *Parameters*. Click on it and define *x: 0*.

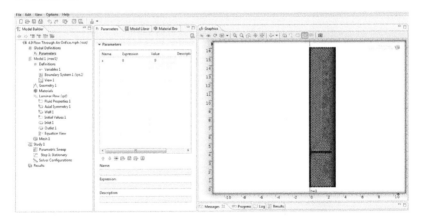

*Step 4*

- Now, right click on the *Laminar Flow* tab in *Model Builder*. Select the *Inlet* option. Another window will open namely *Inlet 1*. Click on it.
- Select the bottom boundary on the rectangle graph. Click on the "+" sign at the top right corner, this will add *inlet boundary* in the *Boundary Selection* tab. At this point, put $u = \left(\dfrac{2}{36}\right) \times \left(1 - \left(\dfrac{r}{3}\right)^2\right)$, $w = 0$

for velocity field. This step will add a boundary condition as given in Equation (4.44).

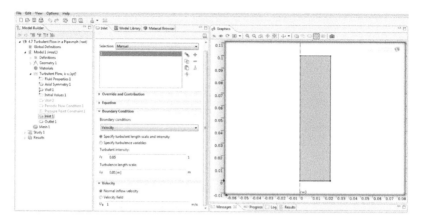

- Again right click on the *Laminar Flow* tab in *Model Builder*. Select the *Outlet* option. Another window will open namely *Outlet 1*. Click on it.
- Select the top boundary on the rectangle graph. Click on the "+" sign at the top right corner, this will add *outlet boundary* in the *Boundary Selection* tab. At this point, put *Pressure $p_0$: 0*. This step will add a boundary condition as given in Equation (4.45).

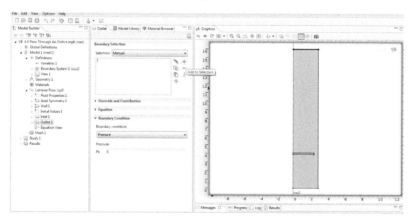

- Now, in the *Model Builder*, select *Wall 1* option available in *Laminar Flow*. At this point, put boundary condition to *No slip*. This step will add a boundary condition as given in Equation (4.47).

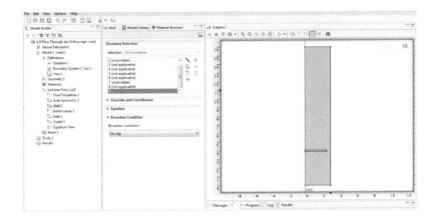

- Now, in the *Model Builder*, select *Axial Symmetry 1* option available in *Laminar Flow*. In COMSOL, by default, the axisymmetric geometry allows slip and this step will add a boundary condition as given in Equation (4.46).

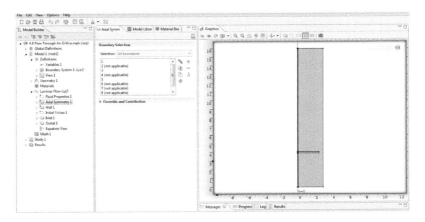

*Step 5*

- Click on the *Mesh* option in *Model Builder*. Select *Normal Mesh* Type. Click on *Build All* option at the top of ribbon. A dialogue box will appear in the *Message* tab as: "*Complete mesh consists of 4518 elements*"

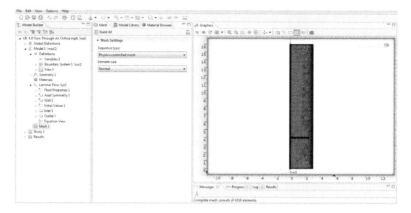

- Now, right click on the *Study 1* tab in *Model Builder*. Select the *Parametric Sweep* option. Another window will open namely *Study Settings*. Click on "+" sign to add "*x*" in the *Parameter names* and set the range (0,0.1,1.5). This will vary the Reynolds number from 1 to 30.

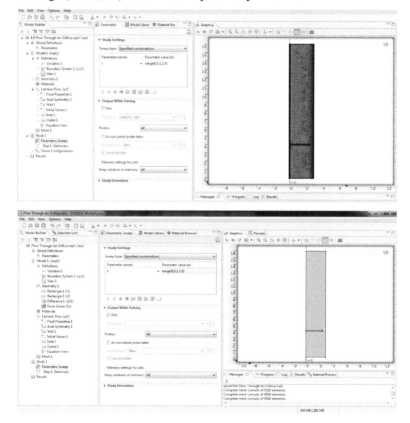

- Now, go to *Study 1* option in the model pellet tab. Click on the *Compute (=)* button. A plot of velocity magnitude, $U = \text{sqrt}(u2 + w2)$ will appear.
- Save the simulation.

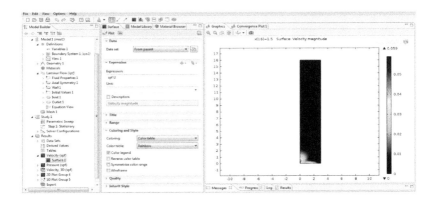

*Step 6*

- In the *Model Builder*, right click on the *Results* tab and choose the *2D Plot Group* option. Right click on it and select *Streamline*. Another window will open namely *Streamline 1*. Figure 4.15 shows the plot for streamlines for flow through an orifice for $R_e = 10^{1.5}$.

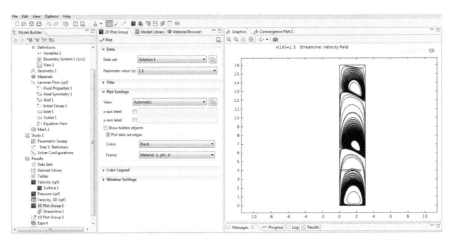

**FIGURE 4.15**    Streamlines for flow through an orifice ($R_e = 10^{1.5}$).

- Now, change the value of $R_e$ to $10^{0.9}$ in *2D Plot Group 1* and click on plot icon. Figure 4.16 shows the plot for streamlines for flow through an orifice for $R_e = 10^{0.9}$.

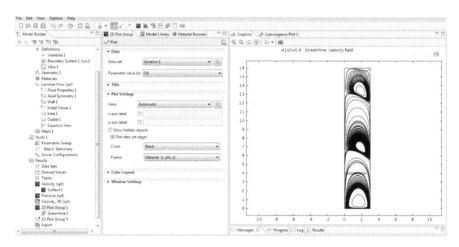

**FIGURE 4.16**   Streamlines for flow through an orifice ($R_e = 10^{0.9}$).

## 4.8  PROBLEMS

1.  Consider a semi-infinite body of liquid bounded by horizontal surface in the *xz* plane (Gupta, 1995). It has constant density and viscosity. It is assumed that the fluid and the solid are at rest initially. The solid surface is then set in motion with velocity $v_0$ at time $t = 0$. It is assumed that the flow is laminar with no pressure gradient in the *x* direction. The governing differential equation for viscous flow of fluid near a wall suddenly set in motion is

$$\frac{\partial v_x}{\partial t} = v \frac{\partial^2 v_x}{\partial y^2}$$

The initial and boundary conditions are
At $t \le 0$, $v_x = 0$ for all $y$
At $y = 0$, $v_x = 1$ for all $t > 0$
At $y = 1$, $v_x = 0$ for all $t > 0$.
Find the velocity $v_x$ as a function of $y$ and $t$.

2. Consider a semi-infinite reservoir through which a jet of fluid emerges from a circular hole and create a net radial inflow as it moves outward. The boundary layer approximation for the time-smoothed equation of change is given by

Continuity: $\dfrac{1}{r}\dfrac{\partial}{\partial r}(rv_r) + \dfrac{\partial v_z}{\partial z} = 0$

Motion: $v_r\dfrac{\partial v_z}{\partial r} + v_z\dfrac{\partial v_z}{\partial z} = \dfrac{1}{r}\dfrac{\partial}{\partial r}\left(r\dfrac{\partial v_z}{\partial r}\right).$

The boundary conditions are
At $r = 0$, $v_r = 0$
At $r = 1$, $\partial v_z / \partial r = 0$
At $z = \infty$, $v_z = 0$.
Find the velocity distribution in the jet and also the amount of fluid crossing each plane of constant $z$.

3. A viscous fluid is in laminar flow in a circular tube of radius $R$ with constant physical properties $(\mu, \rho, k, C_p)$. The temperature distribution in the fluid as a function of $r$ and $z$ is given by

$$\rho C_p v_{z,max}\left[1-\left(\dfrac{r}{R}\right)^2\right]\dfrac{\partial T}{\partial z} = k\left[\dfrac{1}{r}\dfrac{\partial}{\partial r}\left(r\dfrac{\partial T}{\partial r}\right)\right]$$

At $r = 0$, $T = 298\ K$

At $r = 0.2$ m, $k\dfrac{\partial T}{\partial r} = 2$

At $z = 0$, $T = 298\ K$.

Take $\rho = 890\dfrac{kg}{m^3}, C_p = 2\,J/gC, k = 0.180\dfrac{W}{m\ K}, v_{z,max} = 2.1 m/s.$

Determine the temperature distribution in the tube.

# CHAPTER 5

# Heat and Mass Transfer Processes in 2D and 3D

## 5.1 INTRODUCTION

In this chapter, the application of *COMSOL* is discussed to solve problems in heat and mass transfer using *Mathematics* and *Chemical Engineering* modules. The examples discussed here include steady and transient heat transfer in 2D processes, heat conduction problem with a hole at the center of the slab, incorporation of convection–diffusion in microfluidic devices, incorporation of concentration dependent viscosity in microfluidic devices, incorporation of chemical reaction in microfluidic devices, and incorporation of convection–diffusion in a 3D *T*-sensor. The examples demonstrate making different plots such as streamlines, velocity, and so on, to calculate different properties and application of parametric solver.

## 5.2 SIMULATION OF HEAT TRANSFER IN TWO DIMENSIONS

### 5.2.1 PROBLEM STATEMENT

Consider the heat conduction at steady state in a long square slab ($3L3L$), in which heat is generated at a uniform rate of W/m$^3$ (Ghoshdastidar, 1998). The problem is assumed to be 2D, and end effects are neglected. The governing energy equation and boundary conditions are given as follows:

$$\frac{\partial^2 T}{\partial x^2} + \frac{\partial^2 T}{\partial y^2} = 0 \text{ in } 0 \le x \le 1, 0 \le y \le 1 \tag{5.1}$$

$$T = 1 \text{ at } x = 0 \tag{5.2a}$$

$$T = 1 \text{ at } y = 0 \tag{5.2b}$$

$$T = 0 \text{ at } y = 1 \tag{5.2c}$$

$$n.\nabla T \equiv \frac{\partial T}{\partial n} = 0 \text{ at } x = 1. \tag{5.2d}$$

## 5.2.2   SIMULATION APPROACH (METHOD 1)

### Step 1

- Open COMSOL Multiphysics.
- Select *2D Space Dimension* from the list of options. Hit the *next* arrow at the upper right corner.

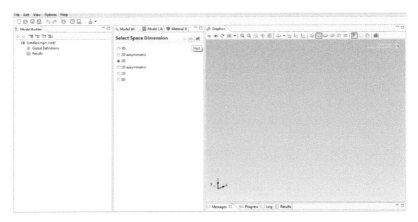

- Select and expand the *Mathematics* folder from the list of options in *Model Wizard*. Further select and expand *PDE Interface* and click on the *Coefficient Form PDE (c)*. Again hit the *next* arrow.

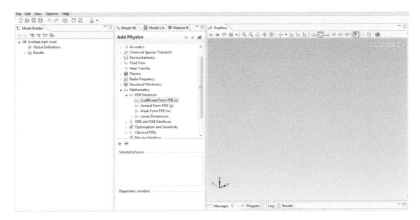

- Then, select *Stationary* from the list of *Study Type* options, and click on the *Finish flag* at the upper right corner of the application.

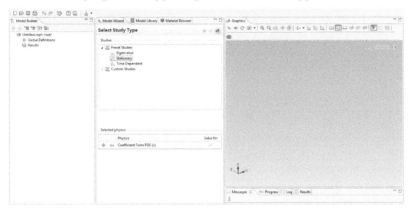

## Step 2

- Select the *Root* node and change the *Unit System: None*.

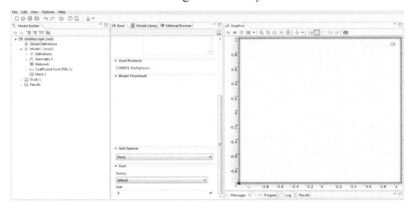

## Step 3

- With the *Model Builder*, right click on the *Geometry* option. From the list of options, click on the *Square*.
- In the *Square 1* environment, choose *Side length:1* with *Base Corner: x = 0, y = 0*. Click on the *Build All* option at the upper part of tool bar. A square graph will appear on the right side of the application window.

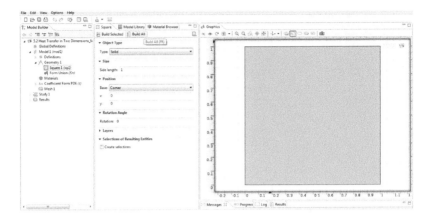

## Step 4

- In the *Model Builder*, select *Coefficient Form PDE* node and change the Dependent variables to T.

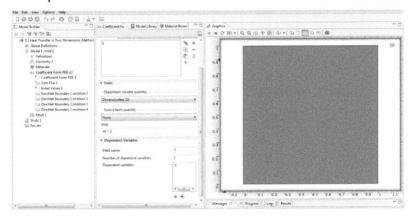

- Now in the *Model Builder*, select *Coefficient Form PDE 1* option. This will open a tab to enter the coefficients of the characteristics of a model equation.
- In the *Domain Selection* panel, you will see an equation of the form

$$e_a \frac{\partial^2 T}{\partial t^2} + d_a \frac{\partial T}{\partial t} + \nabla \cdot \left( -c\nabla T - \alpha T + \gamma \right) + \beta \cdot \nabla T + aT = f \qquad (5.3)$$

where $\nabla = \left[ \dfrac{\partial}{\partial x}, \dfrac{\partial}{\partial y} \right]$.

Assign the coefficients in above equation a suitable value.

To convert Equation (5.3) to the desired form of Equation (5.1), the value of coefficients in Equation (5.3) to be changed as follows:

$c = -1$ $\alpha = 0,$ $\beta = 0,$ $\gamma = 0$

$a = 0 f = 0$

$e_a = 0 d_a = 0.$

This adjustment will reduce Equation (5.3) to Equation (5.1) format.

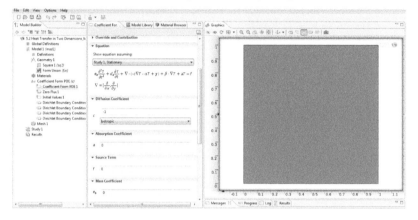

*Step 5*

- Now, right click on the *Coefficient Form PDE* tab in *Model Builder*. Select the *Dirichlet Boundary Condition* option. Another window will open namely *Dirichlet Boundary Condition 1*. Click on it.
- Select the left boundary on the square graph. Click on the "+" sign at the top right corner; this will add *boundary 1* in the *Boundary Selection* tab. At this point, put *r: 1*. This step will add a *Dirichlet boundary condition*: $T = 1$ at $x = 0$ as given in Equation (5.2a) (Note: as $T = r,$ $r = 1,$ $T = 1$).

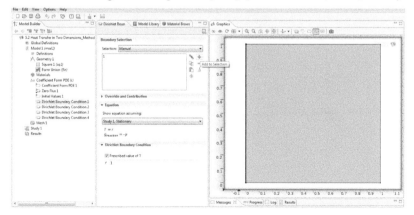

- Again right click on the *Coefficient Form PDE* tab in *Model Builder.* Select the *Dirichlet Boundary Condition* option. Another window will open namely *Dirichlet Boundary Condition 2.* Click on it.
- Select the bottom boundary on the square graph. Click on the "+" sign at the top right corner, this will add *boundary 2* in the *Boundary Selection* tab. At this point, put *r: 1.* This step will add a *Dirichlet boundary condition*: *T = 1 at y = 0* as given in Equation (5.2b) (Note: as *T = r, r = 1, T = 1*).

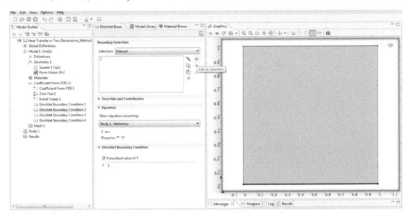

- Again right click on the *Coefficient Form PDE* tab in *Model Builder.* Select the *Dirichlet Boundary Condition* option. Another window will open namely *Dirichlet Boundary Condition 3.* Click on it.
- Select the top boundary on the square graph. Click on the "+" sign at the top right corner, this will add *boundary 3* in the *Boundary Selection* tab. At this point, put *r: 0.* This step will add a *Dirichlet boundary condition*: *T = 0 at y = 1* as given in Equation (5.2c) (Note: as *T = r, r = 0, T = 0*).

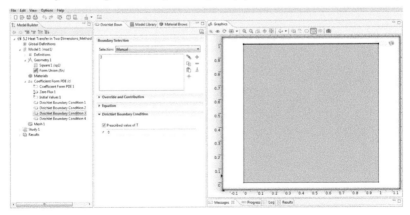

- Again right click on the *Coefficient Form PDE* tab in *Model Builder.* Select the *Dirichlet Boundary Condition* option. Another window will open namely *Dirichlet Boundary Condition 4.* Click on it.
- Select the right boundary on the square graph. Click on the "+" sign at the top right corner, this will add *boundary 4* in the *Boundary Selection* tab. At this point, put *r: T + Tx.* This step will add the *Neumann Boundary Condition* $\dfrac{dT}{dn} = 0$ at $x = 0$, that is, $\dfrac{dT}{dx} = 0$ at $x = 0$, that is, $Tx = 0$ at $x = 0$ as given in Equation (5.2d) (Note: as $T = r,\ r = T + Tx,\ Tx = 0$).

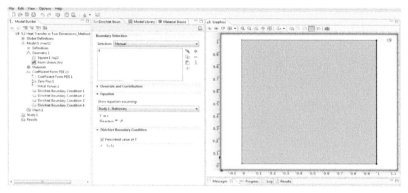

*Step 6*

- Click on the *Mesh* option in *Model Builder.* Select *Normal Mesh* Type. Click on *Build All* option at the top of ribbon. A dialogue box will appear in the *Message* tab as: *"Complete mesh consists of 578 elements"* as shown in Figure 5.1.

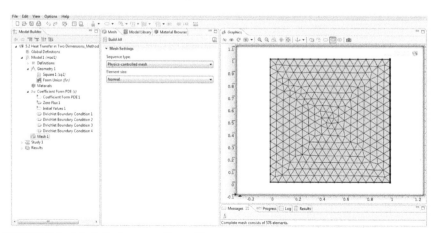

**FIGURE 5.1**    Mesh for heat transfer problem.

- Now, go to *Study* option in the model pellet tab. Click on the *Compute (=)* button. A graph will appear giving the temperature surface plot.
- Save the simulation.

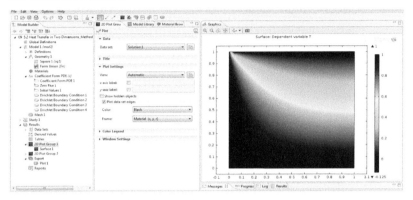

- Choose the *2D Plot Group* option by right clicking on *Results* tab. Another window will open namely *2D Plot Group 2.* Select *Contour* by right clicking on it. Another window will open namely *Contour 1* as shown in Figure 5.2.

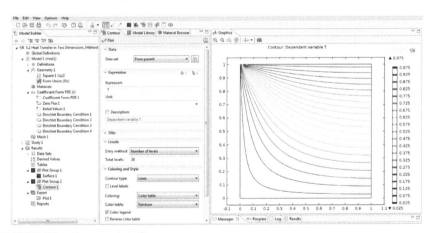

**FIGURE 5.2**    Contour plot of temperature.

## 5.2.3   SIMULATION APPROACH (METHOD 2)

*Step 1*

- Open COMSOL Multiphysics.
- Select *2D Space Dimension* from the list of options. Hit the *next* arrow at the upper right corner.

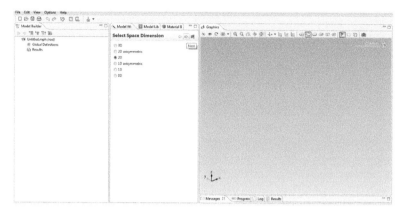

- Select and expand the *Heat Transfer* folder from the list of options in *Model Wizard*. Further select and *Heat Transfer in Solids*. Again hit the *next* arrow.

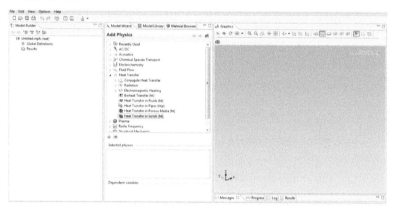

- Then select *Stationary* from the list of *Study Type* options, and click on the *Finish flag* at the upper right corner of the application.

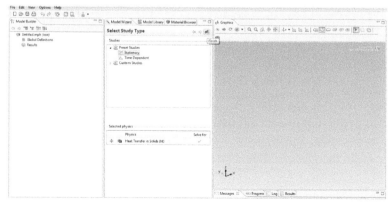

*Step 2*

- Select the *Root* node and change the *Unit System: None*.

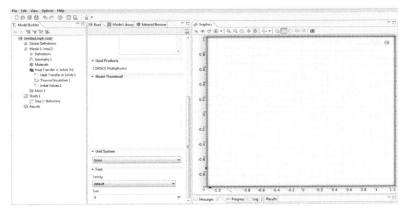

*Step 3*

- With the *Model Builder*, right click on the *Geometry* option. From the list of options, click on the *Square*.
- In the *Square 1* environment, choose *Side length:1* with *Base Corner: x* = 0, *y* = 0. Click on the *Build All* option at the upper part of tool bar. A square graph will appear on the right side of the application window.

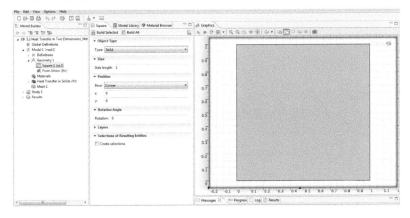

- Now, in the *Model Builder*, select *Heat Transfer in Solids 1* option. This will open a tab to enter the coefficients of the characteristics of a model equation.
- In the *Domain Selection* panel, you will see an equation of the form

$$\rho C_p u \cdot \nabla T = \nabla \cdot \left( k \nabla T \right) + Q \tag{5.4}$$

where $\nabla = \left[ \dfrac{\partial}{\partial x}, \dfrac{\partial}{\partial y} \right]$.

Assign the coefficients in above equation a suitable value.

To convert Equation (5.4) to the desired form of Equation (5.1), the value of coefficients in Equation (5.4) to be changed as follows:

$k = 1$.

This adjustment will reduce Equation (5.4) to Equation (5.1) format.

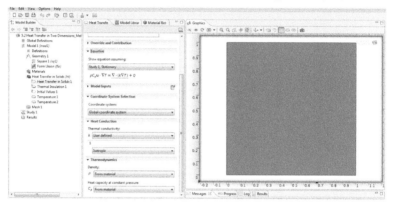

*Step 4*

- In the *Model Builder*, select *Temperature* option by right clicking on the *Heat Transfer in Solids* tab. Another window will open namely *Temperature 1*. Click on it.
- Select the left and bottom boundaries on the square graph. Click on the "+" sign at the top right corner; this will add *boundaries 1 and 2* in the *Boundary Selection* tab. At this point, put $T_0$: 1. This step will add a *Dirichlet boundary conditions* of Equations (5.2a) and (5.2b).

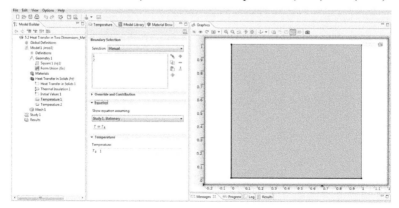

- In the *Model Builder*, select *Temperature* option by right clicking on the *Heat Transfer in Solids* tab. Another window will open namely *Temperature 2*. Click on it.
- Select the top boundary in the square graph. Click on the "+" sign at the top right corner; this will add *boundary 3* in the *Boundary Selection* tab. At this point, put $T_0$: 0. This step will add a *Dirichlet boundary condition*: $T = 0$ at $y = 1$ as given in Equation (5.2c).

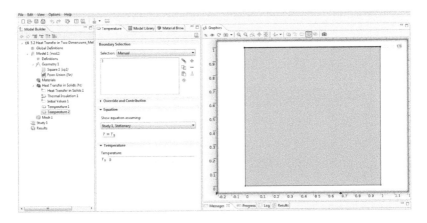

- Now, select *Thermal Insulation 1* option available in the *Heat Transfer in Solids*. Then, select the right boundary on the square graph. The equation at this point is: $-n.(-k\nabla T) = 0$. This is same as the *Neumann Boundary Condition* $\dfrac{dT}{dn} = 0$ at $x = 0$, i.e. $\dfrac{dT}{dx} = 0$ at $x = 0$.

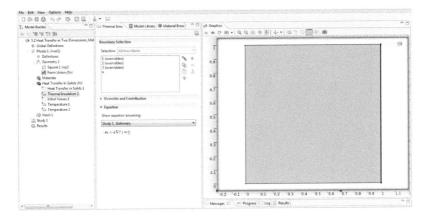

*Step 5*

- Click on the *Mesh* option in *Model Builder.* Select *Normal Mesh* Type. Click on *Build All* option at the top of ribbon. A dialogue box will appear in the *Message* tab as: *"Complete mesh consists of 578 elements"* as shown in Figure 5.3.

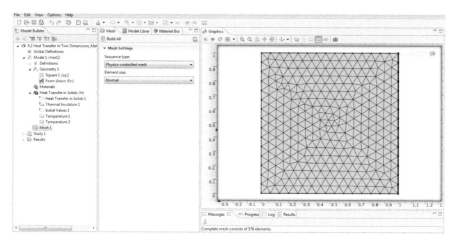

**FIGURE 5.3** Mesh for heat transfer problem.

- Now, go to *Study* option in the model pellet tab. Click on the *Compute* (*=*) button. A graph will appear giving the temperature surface plot.
- Save the simulation.

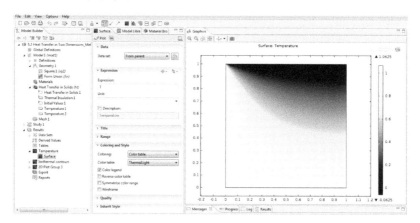

- Choose the *2D Plot Group* option by right clicking on *Results* tab. Another window will open namely *2D Plot Group 2.* Select *Contour*

by right clicking on it. Another window will open namely *Contour 1* as shown in Figure 5.4.

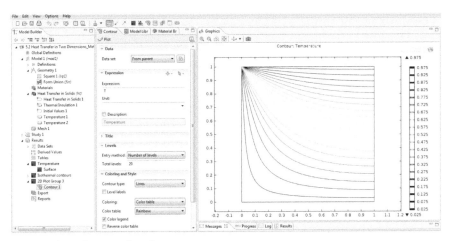

**FIGURE 5.4**    Contour plot of temperature.

## 5.3    SIMULATION OF HEAT CONDUCTION WITH A HOLE

### 5.3.1    *PROBLEM STATEMENT*

Consider the heat conduction at steady state in a long square slab (3L3L) in which heat is generated at a uniform rate of W/m³. The problem is assumed to be 2D, and end effects are neglected. The governing energy equation and boundary conditions are given as follows:

$$\frac{\partial^2 T}{\partial x^2} + \frac{\partial^2 T}{\partial y^2} = 0 \text{ in } 0 \le x \le 1, 0 \le y \le 1 \tag{5.5}$$

$$T = 1 \text{ at } x = 0 \tag{5.6a}$$

$$T = 1 \text{ at } y = 0 \tag{5.6b}$$

$$T = 0 \text{ at } y = 1 \tag{5.6c}$$

$$n.\nabla T \equiv \frac{\partial T}{\partial n} = 0 \text{ at } x = 1. \tag{5.6d}$$

Introduce a hole with side length of 0.4 in the middle of a long square slab (3L3L). Also, check the heat flux on the lower surface of the hole. The

governing energy equation and boundary conditions for the hole are given
as follows:

$$\frac{\partial T}{\partial x} = 0 \text{ at } x = 0 \tag{5.7a}$$

$$\frac{\partial T}{\partial x} = 0 \text{ at } x = 1 \tag{5.7b}$$

$$\frac{\partial T}{\partial y} = 0 \text{ at } y = 0 \tag{5.7c}$$

$$\frac{\partial T}{\partial y} = 0 \text{ at } y = 1. \tag{5.7d}$$

### 5.3.2   SIMULATION APPROACH (METHOD 1)

*Step 1*

- Open the previous solution (Example 1_Method 1).

*Step 2*

- With the *Model Builder*, right click on the *Geometry* option. From the
  list of options, click on the *Square*.
- In the *Square 1* environment, choose *Side length:0.4* with *Base Corner:*
  $x = 0, y = 0$. Click on the *Build All* option at the upper part of tool bar.

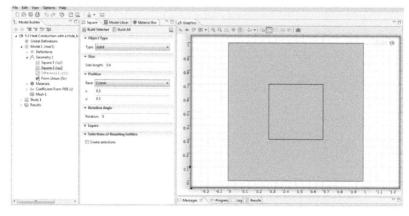

- With the *Model Builder*, again right click on the *Geometry* option.
  From the list of options, click on the *Boolean Operations* and select
  *Difference*.

- In the *Difference 1* environment, add the *Square 1* and subtract the *Square 2*. Click on the *Build All* option at the upper part of tool bar. A square graph with a hole will appear on the right side of the application window.

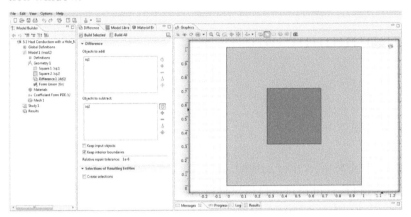

*Step 3*

- Right click on the *Coefficient Form PDE* tab in *Model Builder.* Select the *Dirichlet Boundary Condition* option. Another window will open namely *Dirichlet Boundary Condition 5.* Click on it.
- Select the left boundary on the hole. Click on the "+" sign at the top right corner; this will add *left boundary* in the *Boundary Selection* tab. At this point, put *r: T + Tx.* This step will add the *Neumann Boundary Condition* $\frac{dT}{dx} = 0$ at $x = 0$, that is, $Tx = 0$ at $x = 0$, as given in Equation (5.7a) (Note: as $T = r$, $r = T + Tx$, $Tx = 0$).

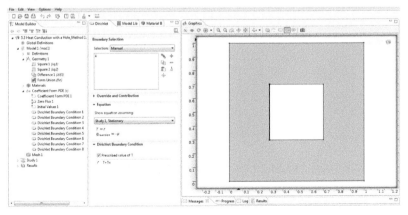

- Again right click on the *Coefficient Form PDE* tab in *Model Builder.* Select the *Dirichlet Boundary Condition* option. Another window will open namely *Dirichlet Boundary Condition 6.* Click on it.
- Select the right boundary on the hole. Click on the "+" sign at the top right corner, this will add *right boundary* in the *Boundary Selection* tab. At this point, put *r: T + Tx.* This step will add the *Neumann Boundary Condition* $\frac{dT}{dx} = 0$ *at x = 1,* that is, *Tx = 0 at x = 1* as given in Equation (5.7b) (Note: as *T = r, r = T + Tx, Tx = 0*).

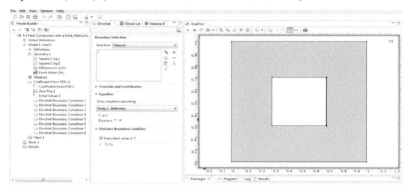

- Again right click on the *Coefficient Form PDE* tab in *Model Builder.* Select the *Dirichlet Boundary Condition* option. Another window will open namely *Dirichlet Boundary Condition 7.* Click on it.
- Select the bottom boundary on the hole. Click on the "+" sign at the top right corner, this will add *bottom boundary* in the *Boundary Selection* tab. At this point, put *r: T + Ty.* This step will add the *Neumann Boundary Condition* $\frac{dT}{dy} = 0$ *at y = 0,* that is, *Ty = 0 at y = 0* as given in Equation (5.7c) (Note: as *T = r, r = T + Ty, Ty = 0*).

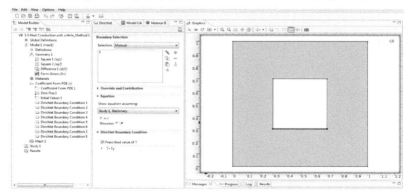

- Again right click on the *Coefficient Form PDE* tab in *Model Builder.* Select the *Dirichlet Boundary Condition* option. Another window will open namely *Dirichlet Boundary Condition 8.* Click on it.
- Select the top boundary on the hole. Click on the "+" sign at the top right corner, this will add *top boundary* in the *Boundary Selection* tab. At this point, put *r: T + Ty.* This step will add the *Neumann Boundary Condition* $\dfrac{dT}{dy} = 0$ *at* $y = 1$ that is, $Ty = 0$ *at* $y = 1$ as given in Equation (5.7d) (Note: as $T = r, r = T + Ty, Ty = 0$).

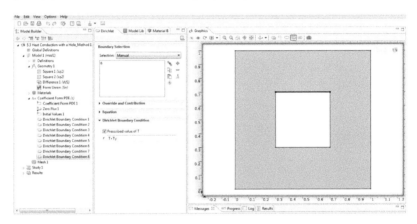

*Step 4*

- Click on the *Mesh* option in *Model Builder.* Select *Normal Mesh* Type. Click on *Build All* option at the top of ribbon. A dialogue box will appear in the *Message* tab as: *"Complete mesh consists of 572 elements."*

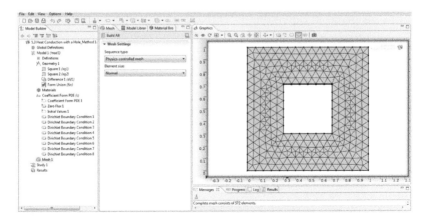

- Now, go to *Study* option in the model pellet tab. Click on the *Compute* (=) button. A graph will appear giving the temperature surface plot.
- Save the simulation.

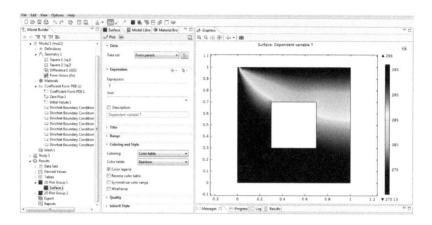

- Choose the *2D Plot Group* option by right clicking on *Results* tab. Another window will open namely *2D Plot Group 2*. Select *Contour* by right clicking on it. Another window will open namely *Contour 1* as shown in Figure 5.5.

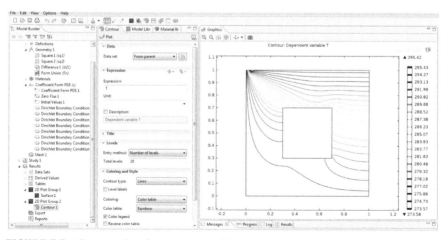

**FIGURE 5.5**  Contour plot of temperature with a hole.

*Step 5*

- To calculate the heat flux on the lower surface of the hole, select *Derived Values* option available in the *Results* tab.
- Now, right click on the *Derived Values* tab and select the *Line Integration* option. Another window will open namely *Line Integration 1*. Click on it. In the Expression section set *Expression: Ty*. Select the bottom boundary of the hole. Click on the "+" sign at the top right corner; this will add *bottom boundary* in the *Boundary Selection* tab.
- Click on "= Evaluate" button at the top. It is around $8.21914 \times 10^{-13}$, which is close to zero as given by Equation (5.7c).

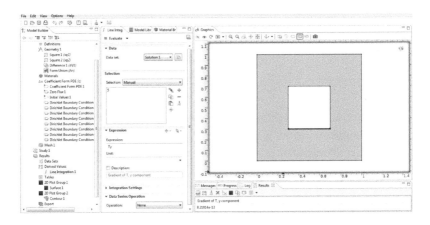

### 5.3.3  *SIMULATION APPROACH (METHOD 2)*

*Step 1*

- Open the previous solution (Example 1_Method 2).

*Step 2*

- With the *Model Builder*, right click on the *Geometry* option. From the list of options, click on the *Square*.
- In the *Square 1* environment, choose *Side length:0.4* with *Base Corner: $x = 0$, $y = 0$*. Click on the *Build All* option at the upper part of tool bar.

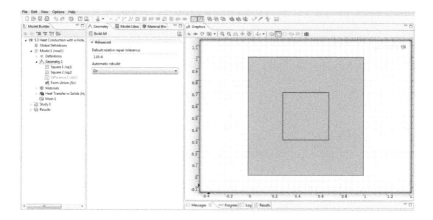

- With the *Model Builder*, again right click on the *Geometry* option. From the list of options, click on the *Boolean Operations* and select *Difference*.
- In the *Difference 1* environment, add the *Square 1* and subtract the *Square 2*. Click on the *Build All* option at the upper part of tool bar. A square graph with a hole will appear on the right side of the application window.

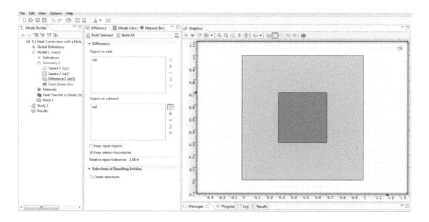

*Step 3*

- Now, select *Thermal Insulation 1* option available in the *Heat Transfer in Solids*. All the hole boundaries have the equation: $-n.(-k\nabla T) = 0$.

This is same as the *Neumann Boundary Conditions* given by Equations (5.7a)–(5.7d).

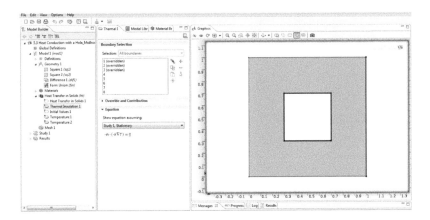

## Step 4

- Click on the *Mesh* option in *Model Builder*. Select *Normal Mesh* Type. Click on *Build All* option at the top of ribbon. A dialogue box will appear in the *Message* tab as: *"Complete mesh consists of 572 elements."*

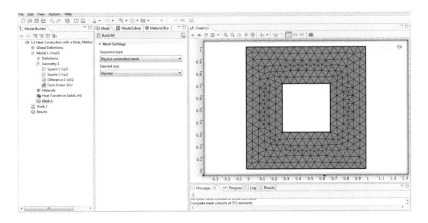

- Now, go to *Study* option in the model pellet tab. Click on the *Compute* (=) button. A graph will appear giving the temperature surface plot.
- Save the simulation.

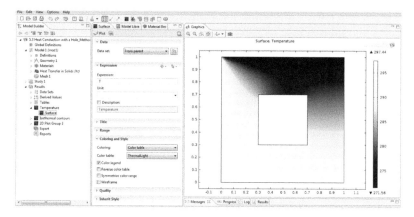

- Choose the *2D Plot Group* option by right clicking on *Results* tab. Another window will open namely *2D Plot Group 2*. Select *Contour* by right clicking on it. Another window will open namely *Contour 1* as shown in Figure 5.6.

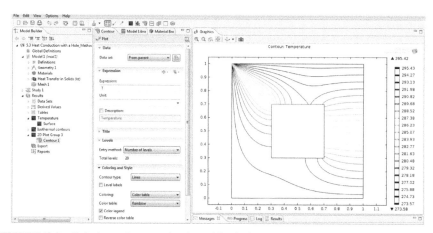

**FIGURE 5.6** Solution to heat conduction with a hole.

*Step 5*

- To calculate the heat flux on the lower surface of the hole, select *Derived Values* option available in the *Results* tab.
- Now, right click on the *Derived Values* tab and select the *Line Integration* option. Another window will open namely *Line Integration 1*. Click on it. In the Expression section set *Expression: Ty*. Select the bottom boundary of the hole. Click on the "+" sign at the top right corner, this will add *bottom boundary* in the *Boundary Selection* tab.

- Click on "= Evaluate" button at the top. It is around −0.45583. It should be close to zero.

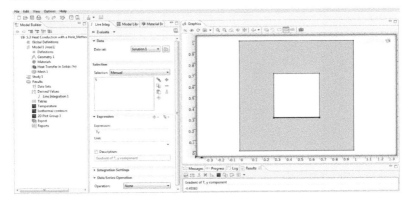

- Click on the *Mesh* option in *Model Builder*. Select *Normal Mesh* Type. Click on *Build All* option at the top of ribbon.
- Now, go to *Study* option in the model pellet tab. Click on the *Compute* (=) button.
- Click on "= Evaluate" button at the top. It is around −0.12396, which is close to zero as given by Equation (5.7c).

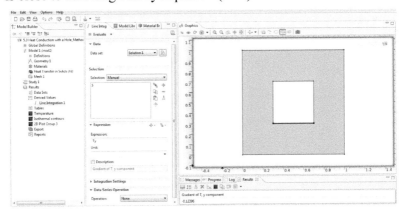

## 5.4  SIMULATION INCORPORATING CONVECTION DIFFUSION IN MICROFLUIDIC DEVICES

### 5.4.1  PROBLEM STATEMENT

The T-sensor discussed in Chapter 4 is incorporated with the convective diffusion equation (Finlayson, 2006). This equation helps in incorporating the transfer of a chemical from one flowing stream to the other. The convection–diffusion equation at steady-state transport processes is

$$\nabla \cdot \left(-D_i \nabla c_i\right) + u \cdot \nabla c_i = R_i. \tag{5.8}$$

In this equation, $c_i$, $D_i$, and $R_i$ are concentration (mol/m$^3$), diffusivity (m$^2$/s), and rate expression for species $i$ (mol/(m$^3$ s)), respectively.

Boundary conditions:

$$\text{At } y = 1.5, c_{0,c} = 0 \text{ at any } x. \tag{5.9a}$$

Inflow from the bottom:

$$\text{At } y = 0, c_{0,c} = 1 \text{ at any } x. \tag{5.9b}$$

At the outlet, it is assumed that convective mass transport is dominant:

$$\nabla \cdot \left(-D_i \nabla c_i\right) = 0. \tag{5.9c}$$

### 5.4.2 SIMULATION APPROACH

*Step 1*

- Open T-sensor problem file of COMSOL Multiphysics discussed in Chapter 4.
- Right click on *Model 1* and press *Add Physics* tab. In the *Add Physics* tab, expand the *Chemical Species Transport* folder from the list of options and select *Transport of Diluted Species*. Hit the *next* arrow at the upper right corner.

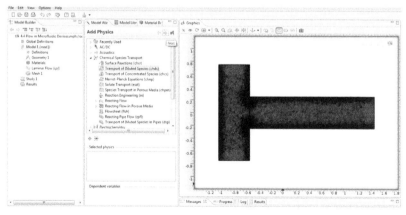

- Select *Stationary* from the list of *Study Type* options, and click on the *Finish flag* at the upper right corner of the application.

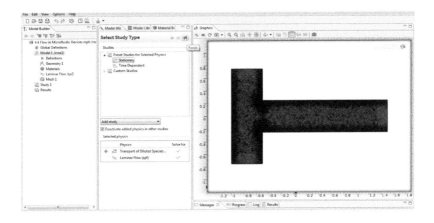

### Step 2

- In the *Model Builder*, select the *Transport of Diluted Species* tab and expand the Dependent variables tab. Change the species concentration to "*c*." Choose the *Discretization tab* and change the *Concentration* from *Linear* to *Quadratic*.

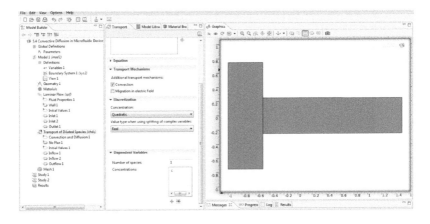

- Now in the *Model Builder*, select *Convection and Diffusion 1* option available in the *Transport of Diluted Species* option. This will open a tab to enter the coefficients of the characteristics of a model equation.
- In the *Domain Selection* panel, you will see an equation of the form

$$\nabla \cdot \left(-D_i \nabla c_i\right) + u \cdot \nabla c_i = R_i \qquad (5.10)$$

$$N_i = -D_i \nabla c_i + u c_i.$$

Assign the coefficients in above equation a suitable value.
Set the Velocity as User defined: *u* and *v*. The Diffusion coefficient to User defined: $1/P_e$.

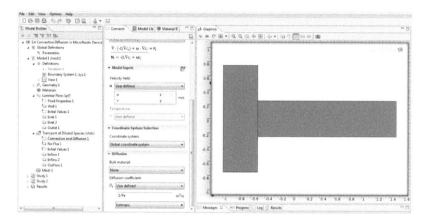

- With the *Model Builder*, right click on the *Definitions* tab and choose *Variables*. Another window will open namely *Variables 1*. Click on it and define $P_e$: *1*.

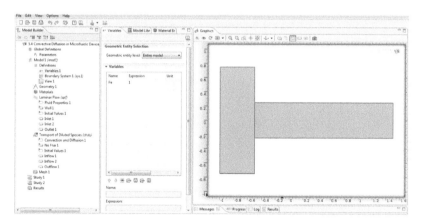

*Step 4*

- Select the *Inflow* option by right clicking on the *Transport of Diluted Species* tab. Another window will open namely *Inflow 1*. Click on it.
- Select the top boundary on the Rectangle 2 graph. Click on the "+" sign at the top right corner, this will add *boundary 3* in the *Boundary*

*Selection* tab. At this point, put *Concentration* $c_{0,c}$: *0*. This step will add a boundary condition as given in Equation (5.9a).

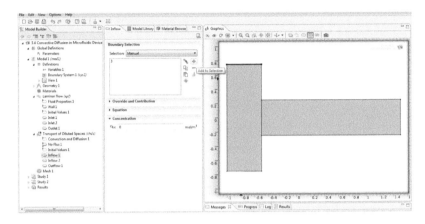

- Again select the *Inflow* option by right clicking on the *Transport of Diluted Species* tab. Another window will open namely *Inflow 2*. Click on it.
- Select the bottom boundary on the Rectangle 2 graph. Click on the "+" sign at the top right corner, this will add *boundary 2* in the *Boundary Selection* tab. At this point, put *Concentration* $c_{0,c}$: *1*. This step will add a boundary condition as given in Equation (5.9b).

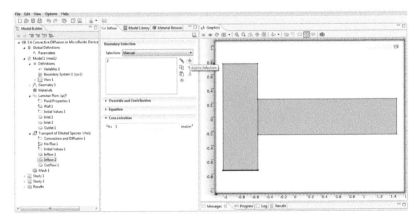

- Again select the *Outflow* option by right clicking on the *Transport of Diluted Species* tab. Another window will open namely *Outflow 1*. Click on it.

- Select the right boundary on the Rectangle 1 graph. Click on the "+" sign at the top right corner, this will add *boundary 9* in the *Boundary Selection* tab. This will add the equation $-n.D_i\nabla_{ci} = 0$ which is same as the outlet boundary condition as given in Equation (5.9c).

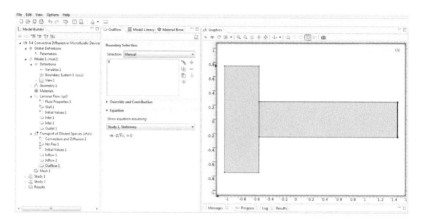

*Step 5*

- Click on the *Mesh* option in *Model Builder*. Select *Normal Mesh* Type. Click on *Build All* option at the top of ribbon. A dialogue box will appear in the *Message* tab as: *"Complete mesh consists of 2101 elements."*

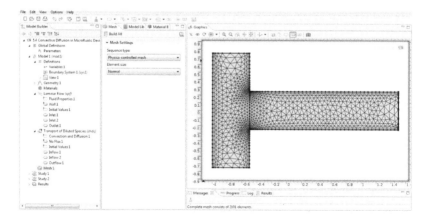

- Now, go to *Study 2* option in the model pellet tab. Click on the *Compute* (=) button. A graph will appear giving the velocity surface plot.
- Save the simulation.

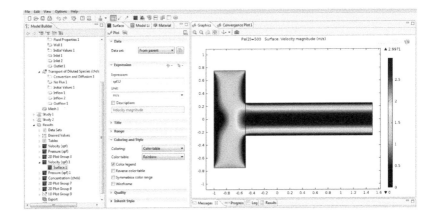

- To see the concentration, change the expression spf. *U* to *c* and click on plot icon.

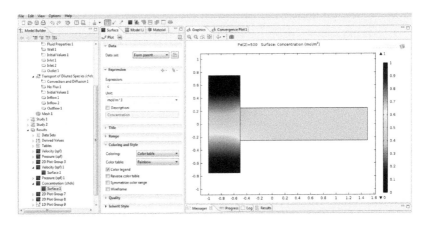

- Choose the *2D Plot Group* option by right clicking on *Results* tab. Another window will open namely *2D Plot Group 2*. Select *Contour* by right clicking on it. Another window will open namely *Contour 1* as shown in Figure 5.7.

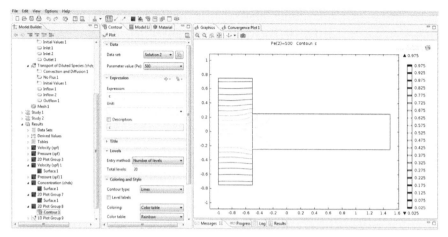

**FIGURE 5.7** Contour plot of convection diffusion problem with $D = 1$.

*Step 6*

- To calculate the overall mass balance, select *Derived Values* option available in the *Results* tab.
- Select the *Line Integration* option by right clicking on *Derived Values* tab.
- Integrate $v*c$ over the top. Select the top inlet boundary on the Rectangle 2 graph. Click on the "+" sign at the top right corner, this will add *boundary 3* in the *Boundary Selection* tab. Click on "= Evaluate" button at the top. The velocity magnitude is 0.

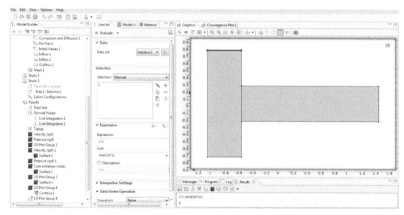

- Now select the bottom inlet boundary on the Rectangle 2 graph. Click on the "+" sign at the top right corner, this will add *boundary 2* in the *Boundary Selection* tab. Click on "= Evaluate" button at the top. The velocity magnitude is 0.49854.

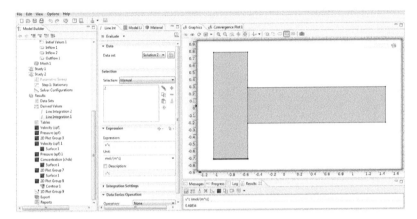

- To integrate *u*c* at the outlet, select the right outlet boundary on the Rectangle 1 graph. Click on the "+" sign at the top right corner, this will add *boundary 9* in the *Boundary Selection* tab. Click on "= Evaluate" button at the top. The velocity magnitude is 0.49856 and mass balance found matches well.

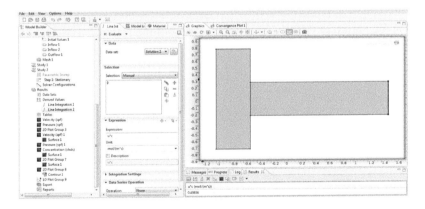

- Now change the Peclet number from 1 to 50 in the variables node. Now go to *Study 2* option in the model pellet tab. Click on the *Compute (=)* button. The contour plot is as shown in Figure 5.8. The flow rates in and out are 0.49854 and 0.49854, respectively, which gives a quite accurate.

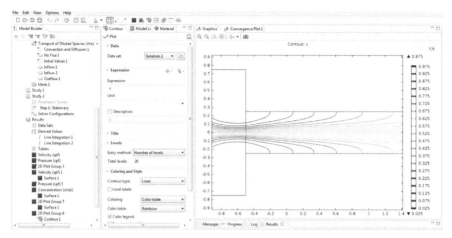

**FIGURE 5.8** Contour plot of convection diffusion problem with $D = 1/50$.

*Step 7*

- Select *Parameters* by right clicking on the *Global Definitions* tab. Another window will open namely *Parameters*. Click on it and define $P_e$: 50. Then, disable *Variables 1* under *Definitions*.
- Now, right click on the *Study 2* tab in *Model Builder*. Select the *Parametric Sweep* option. Another window will open namely *Study Settings*. Click on "+" sign to add "$P_e$" in the *Parameter names* and set the range (50,450,500).

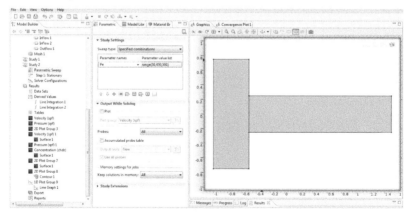

- Now, go to *Study 2* option in the model pellet tab. Click on the *Compute* (=) button.

- Choose the *1D Plot Group* option by right clicking on *Results* tab. Another window will open namely *1D Plot Group 2*. Select *Line graph* by right clicking on it to see the concentration profiles for $P_e = 50$ and $P_e = 500$ as shown in Figure 5.9.

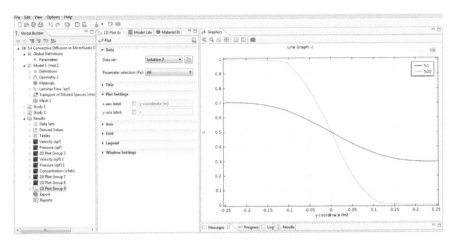

**FIGURE 5.9**   Line graph of exit concentrations for convection diffusion problem with $P_e = 50$ and $P_e = 500$ ($D = 1/50$ and $D = 1/500$).

## 5.5   SIMULATION OF CONCENTRATION-DEPENDENT VISCOSITY

### 5.5.1   *PROBLEM STATEMENT*

Incorporate the concentration-dependent viscosity in the above problem

$$\mu = 1 + c. \tag{5.11}$$

### 5.5.2   *SIMULATION APPROACH*

*Step 1*

- Open file for the "Convection–Diffusion in Microfluidic Devices problem" as discussed above.
- Choose *Parameters* by right clicking on the *Global Definitions* tab. Another window will open namely *Parameters*. Click on it and define *alpha:* 0.

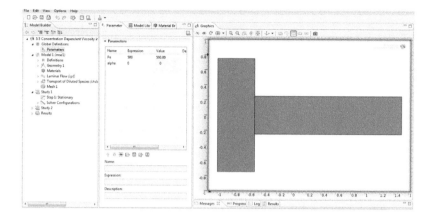

## Step 2

- In the *Model Builder*, select *Fluid Properties 1* option available in the *Laminar Flow* option. Change the viscosity to 1 + alpha*c.

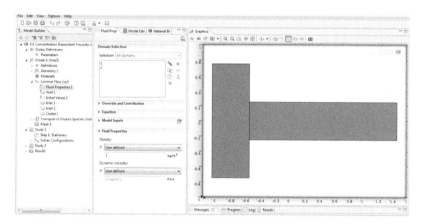

## Step 3

- Now, right click on the *Study 2* tab in *Model Builder*. Select the *Parametric Sweep* option. Another window will open namely *Study Settings*. Click on "+" sign to add "alpha" in the *Parameter names* and set the range (0,1).

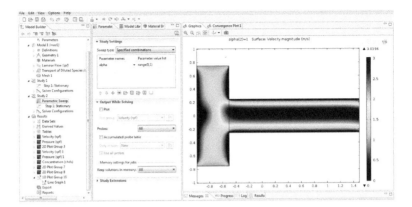

- Now, go to *Study 2* option in the model pellet tab. Click on the *Compute* (=) button.
- Choose the *1D Plot Group* option by right clicking on *Results* tab. Another window will open namely *1D Plot Group 2*. Select *Line graph* by right clicking on it to see the concentration profiles with viscosity = 1 the other $1 + \text{alpha}*c$ for $P_e = 500$ as shown in Figure 5.10.

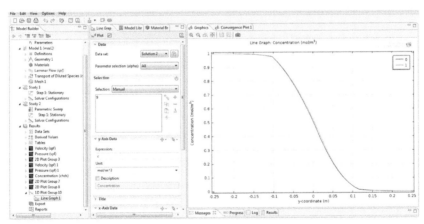

**FIGURE 5.10**   Concentration profiles of *T*-sensor with different viscosities for $D = 1/500$.

## 5.6   SIMULATION OF CHEMICAL REACTION IN MICROFLUIDIC DEVICE

### 5.6.1   PROBLEM STATEMENT

Incorporate the reaction rate in the convection diffusion problem illustrated in Chapter 4.

$$Rate = kc^2, \text{ where rate is in mole per volume per time} \quad (5.12)$$

## 5.6.2 SIMULATION APPROACH

*Step 1*

- Open COMSOL Multiphysics file for flow and diffusion problem discussed in Chapter 4.

*Step 2*

- Select *Reactions1* option available in the *Transport of Diluted Species* tab. Expand the Reactions tab and set the $R_c$: *rate*.

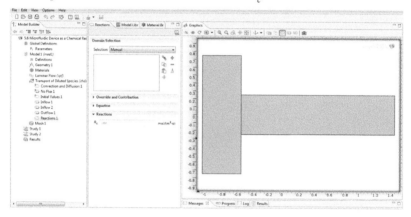

- Choose *Variables* by right clicking on the *Definitions* tab. Another window will open namely *Variables 1*. Click on it and define rate = $-k*c*c$.

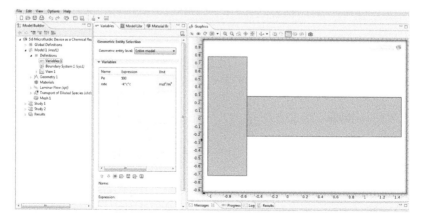

- Select *Parameters* by right clicking on the *Global Definitions* tab. Another window will open namely *Parameters*. Click on it and define *k: 0.1*.

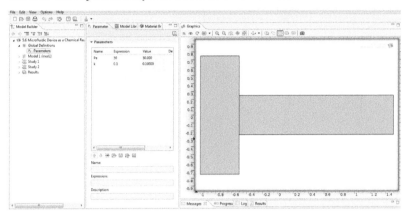

Now, go to *Study 2* option in the model pellet tab. Click on the *Compute* (=) button. The material reacts very quickly.

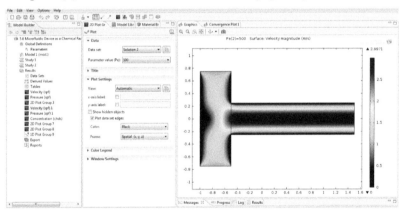

## 5.7 SIMULATION INCORPORATING CONVECTION AND DIFFUSION IN A THREE-DIMENSIONAL *T*-SENSOR

### 5.7.1 *PROBLEM STATEMENT*

Consider the flow through *T* represented by two rectangles with the following dimensions:

Rectangle 1: Width = 2, Depth = 0.5, Height = 0.5,
and Base corner: $x = 0, y = 0, z = 0$.

Rectangle 2: Width = 0.5, Depth = 0.5, Height = 1.5,
and Base corner: $x = 0$, $y = 0$, $z = -0.5$.

It is described mathematically by the Navier–Stokes equation (5.14) and the continuity equation (5.15) in the dimensionless form

$$\rho\left(u\frac{\partial u}{\partial x} + v\frac{\partial u}{\partial y} + w\frac{\partial u}{\partial y}\right) = -\frac{\partial p}{\partial x} + \mu\left(\frac{\partial^2 u}{\partial x^2} + \frac{\partial^2 u}{\partial y^2} + \frac{\partial^2 u}{\partial z^2}\right) \qquad (5.13)$$

$$\frac{\partial u}{\partial x} + \frac{\partial v}{\partial y} + \frac{\partial w}{\partial z} = 0. \qquad (5.14)$$

Boundary conditions:
Inflow from the top:

$$\text{At } z = 1, u = v = w = 1 \text{ at any } x, y. \qquad (5.15)$$

Inflow from the bottom:

$$\text{At } z = -0.5, u = v = w = 1 \text{ at any } x, y. \qquad (5.16)$$

At the outflow the viscous stress in fully developed flow is considered, hence pressure is zero.

$$\text{At } x = 2, p_0 = 0 \text{ } at \text{ } any \text{ } y, z \qquad (5.17)$$

The wall at all other boundaries is having no slip condition with the velocity zero for both components.

Perform the overall mass balance.

Given $\rho = 1$, $\mu = 1$.

*Step 1*

- Open COMSOL Multiphysics.
- Select *2D Space Dimension* from the list of options. Hit the *next* arrow at the upper right corner.

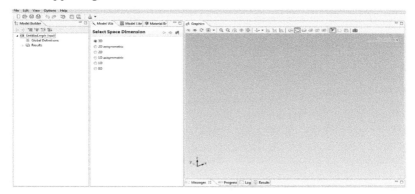

- Select and expand the *Fluid Flow* folder from the list of options in *Model Wizard* and click on the *Single Phase Flow* and select the *Laminar Flow*. Again hit the *next* arrow.

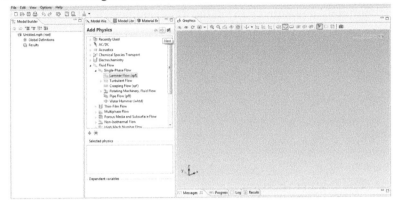

- Then, select *Stationary* from the list of *Study Type* options, and click on the *Finish flag* at the upper right corner of the application.

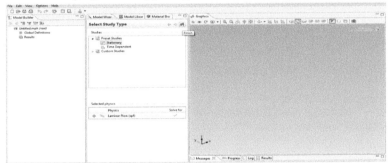

- In the *Model Builder*, select *Root* node and change the Unit System to None.

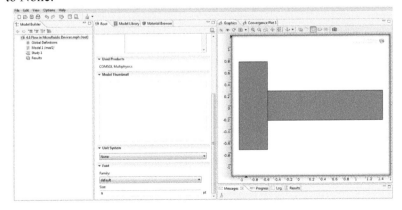

*Step 2*

- In the *Model Builder*, right click on the *Geometry* option. From the list of options, click on the *Rectangle* taskbar.
- In the *Rectangle 1* environment, select *Size* with *Width: 2, Depth: 0.5, Height: 0.5, Base Corner: x* = 0, *y* = 0, *z* = 0. Click on the *Build All* option at the upper part of tool bar. A rectangle graph will appear on the right side of the application window.

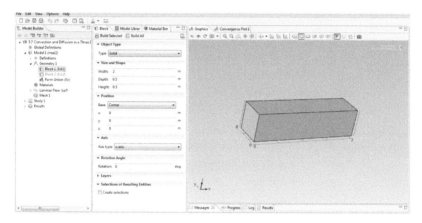

- In the *Model Builder* again, right click on the *Geometry* option. From the list of options, click on the *Rectangle* taskbar.
- In the *Rectangle 2* environment, select *Size* with *Width: 0.5, Depth: 0.5, Height: 1.5, Base Corner: x* = 0, *y* = 0, *z* = –0.5. Click on the *Build All* option at the upper part of tool bar. A rectangle graph will appear on the right side of the application window.

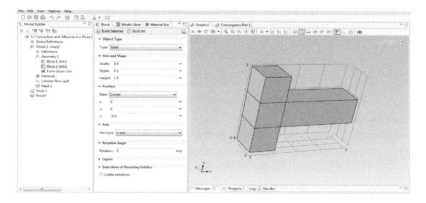

*Step 3*

- Now, in the *Model Builder*, select *Fluid Properties 1* option available in *Laminar Flow*. This will open a tab to enter the coefficients of the characteristics of a model equation.
- In the *Domain Selection* panel, you will see an equation of the form

$$\rho(u.\nabla)u = \nabla.\left[-pl + \mu\left(\nabla u + (\nabla u)^T - \frac{2}{3}\mu(\nabla.u)l\right)\right] + F \qquad (5.18)$$

$$\nabla.(\rho u) = 0 \qquad (5.19)$$

Assign the coefficients in above equation a suitable value.

To convert Equations (5.18) and (5.19) to the desired form of Equations (5.13) and (5.14) respectively, the value of coefficients in Equations (5.18) and (5.19) to be changed as follows:

$\rho = 10, \mu = 1$.

Then, click on the "+" sign at the top right corner, this will add *Domains 1,2,3,4* in the *Domain Selection* tab.

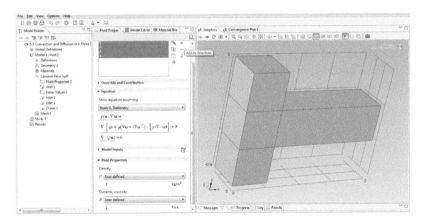

*Step 4*

- Now, right click on the *Laminar Flow* tab in *Model Builder*. Select the *Inlet* option. Another window will open namely *Inlet 1*. Click on it.
- Select the top boundary on the Rectangle 2 graph. Click on the "+" sign at the top right corner, this will add *boundary* in the *Boundary Selection* tab. At this point, put *Normal inflow velocity U₀: 1*. This step will add a boundary condition as given in Equation (5.15).

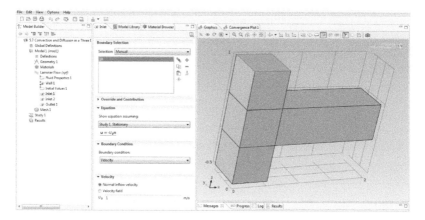

- Again right click on the *Laminar Flow* tab in *Model Builder*. Select the *Inlet* option. Another window will open namely *Inlet 2*. Click on it.
- Select the bottom boundary on the Rectangle 2 graph. Click on the "+" sign at the top right corner, this will add *boundary* in the *Boundary Selection* tab. At this point, put *Normal inflow velocity $U_0$: 1*. This step will add a boundary condition as given in Equation (5.16).

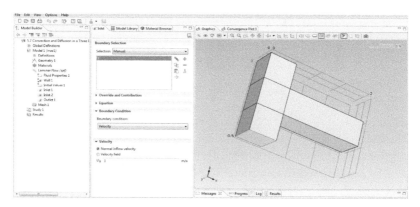

- Again right click on the *Laminar Flow* tab in *Model Builder*. Select the *Outlet* option. Another window will open namely *Outlet 1*. Click on it.
- Select the right boundary on the Rectangle 1 graph. Click on the "+" sign at the top right corner, this will add *boundary* in the *Boundary Selection* tab. At this point, put *Pressure $p_0$: 0*. This step will add a boundary condition as given in Equation (5.17).

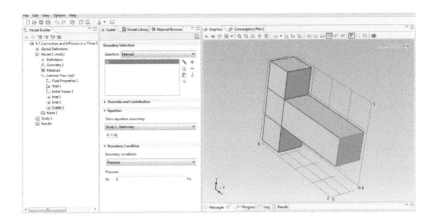

- Now, in the *Model Builder*, select *Wall 1* option available in *Laminar Flow*. By default, it has all other boundaries in the *Boundary Selection* tab. At this point, put boundary condition to *No slip*.

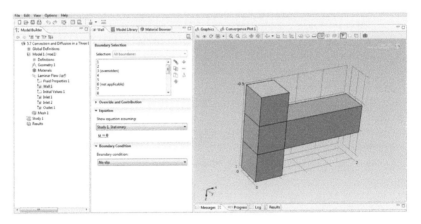

*Step 5*

- Click on the *Mesh* option in *Model Builder*. Select *Normal Mesh* Type. Click on *Build All* option at the top of ribbon. A dialogue box will appear in the *Message* tab as: "*Complete mesh consists of 99131 elements.*"

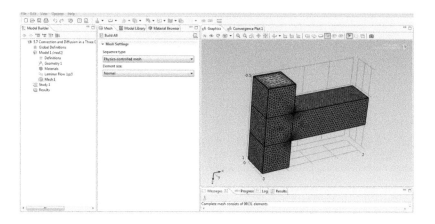

- Now, go to *Study 1* option in the model pellet tab. Click on the *Compute* (=) button.
- Save the simulation.

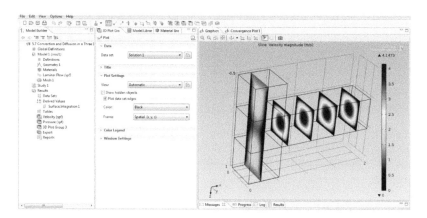

*Step 6*

- In the *Model Builder*, right click on the *Results* tab and select the *3D Plot Group* option. Another window will open namely *3D Plot Group 1*. Right click on it and select *Streamline*. Another window will open namely *Streamline 1*. A plot for streamlines for flow in a *T*-sensor appears as shown in Figure 5.11.

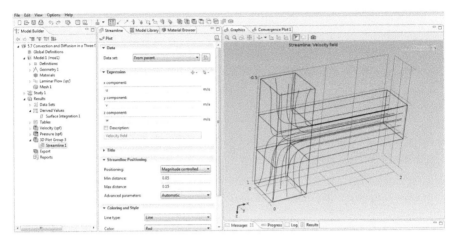

**FIGURE 5.11**   Streamlines of velocity field.

*Step 7*

- To calculate the overall mass balance, select *Derived Values* option available in the *Results* tab.
- Now, right click on the *Derived Values* tab and select the *Surface Integration* option. Another window will open namely *Surface Integration 1*. Click on it.
- Select the top inlet boundary on the Rectangle 2 graph. Click on the "+" sign at the top right corner, this will add *boundary 3* in the *Boundary Selection* tab. Click on "= Evaluate" button at the top. The velocity magnitude is 0.24466.

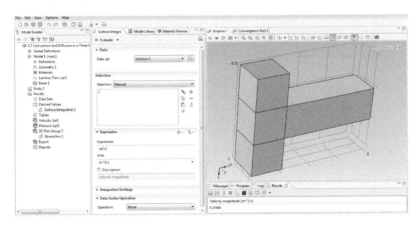

- Now, select the bottom inlet boundary on the Rectangle 2 graph. Click on the "+" sign at the top right corner, this will add *boundary 10* in the *Boundary Selection* tab. Click on "= Evaluate" button at the top. The velocity magnitude is 0.24478.

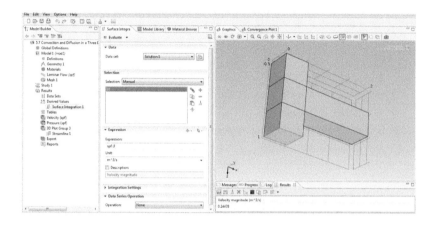

- Now, select the right outlet boundary on the Rectangle 1 graph. Click on the "+" sign at the top right corner, this will add *boundary 21* in the *Boundary Selection* tab. Click on "= Evaluate" button at the top. The velocity magnitude is 0.48411.
- The sum of inlet flow rates is 0.48944 (0.24466 + 0.24478 = 0.48944) and 0.48411 out. This is accurate to within 1.09%.

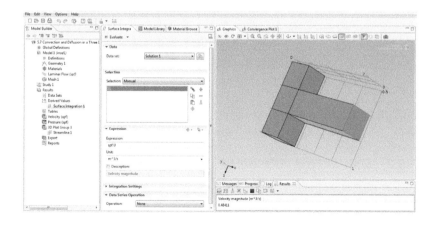

## 5.8   SIMULATION OF TRANSIENT HEAT TRANSFER IN TWO DIMENSIONS

### 5.8.1   *PROBLEM STATEMENT*

The heat transfer is given by the equation

$$\frac{\partial T}{\partial t} = \alpha \left( \frac{\partial^2 T}{\partial x^2} + \frac{\partial^2 T}{\partial y^2} \right) \tag{5.20}$$

The boundary conditions of the slab are shown in Figure 5.12.

The size of the slab is 2 m × 2 m, which is initially at 0 °C. Take $\Delta x = \Delta y = 0.1$ m and $a = 1$ m²/s. The temperature of the slab is to be determined at $t = 0.5$ s. Take $\Delta t = 0.05$ s.

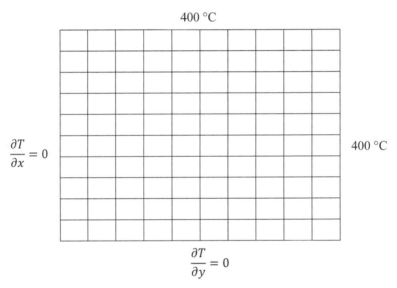

**FIGURE 5.12**   Computational domain of transient heat transfer in two dimensions.

### 5.8.2   *SIMULATION APPROACH*

*Step 1*

- Open COMSOL Multiphysics.
- Select *2D Space Dimension* from the list of options. Hit the *next* arrow at the upper right corner.

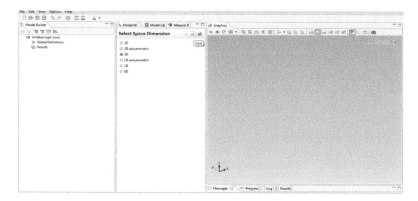

- Select and expand the *Mathematics* folder from the list of options in *Model Wizard*. Further select and expand *PDE Interface* and click on the *Coefficient Form PDE (c)*. Again hit the *next* arrow.

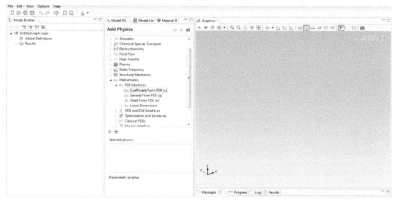

- Then, select *Time Dependent* from the list of *Study Type* options, and click on the *Finish flag* at the upper right corner of the application.

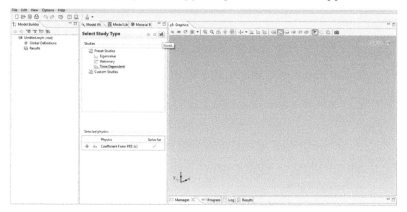

*Step 2*

- With the *Model Builder*, right click on the *Geometry* option. From the list of options, click on the *Square*.
- In the *Square 1* environment, choose *Side length: 2* with *Base Corner: x* = 0, *y* = 0. Click on the *Build All* option at the upper part of tool bar. A square graph will appear on the right side of the application window.

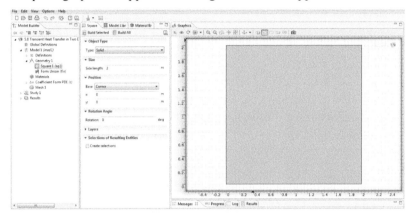

*Step 3*

- In the *Model Builder*, select *Coefficient Form PDE* node and change the dependent variables to *T*.

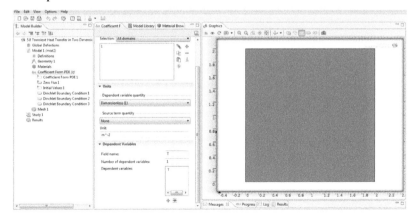

- Now, in the *Model Builder*, select *Coefficient Form PDE 1* option. This will open a tab to enter the coefficients of the characteristics of a model equation.

- In the *Domain Selection* panel, you will see an equation of the form

$$e_a \frac{\partial^2 T}{\partial t^2} + d_a \frac{\partial T}{\partial t} + \nabla.\left(-c\nabla T - \alpha T + \gamma\right) + \beta.\nabla T + aT = f \qquad (5.21)$$

where $\nabla = \left[\dfrac{\partial}{\partial x}, \dfrac{\partial}{\partial y}\right]$.

Assign the coefficients in above equation a suitable value.

To convert Equation (5.21) to the desired form of Equation (5.20), the value of coefficients in Equation (5.21) to be changed as follows:

$c = 1$ $\alpha = 0$, $\beta = 0$, $\gamma = 0$

$a = 0$ $f = 0$

$e_a = 0$ $d_a = 1$.

This adjustment will reduce Equation (5.21) to Equation (5.20) format.

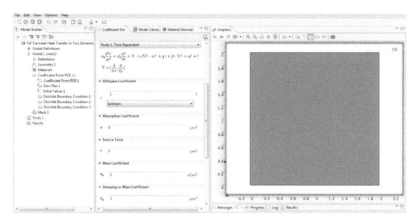

*Step 4*

- Now, right click on the *Coefficient Form PDE* tab in *Model Builder*. Select the *Dirichlet Boundary Condition* option. Another window will open namely *Dirichlet Boundary Condition 1*. Click on it.
- Select the left boundary on the square graph. Click on the "+" sign at the top right corner, this will add *boundary 1* in the *Boundary Selection* tab. At this point, put *r: T + Tx*. This step will add a *Neumann boundary condition* for the left boundary as given in Figure 5.12 (Note: as $T = r$, $r = T + Tx$, $T = Tx$).

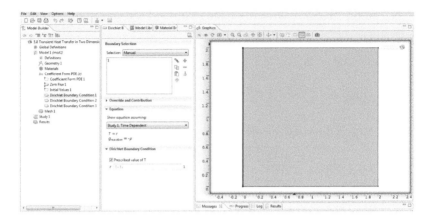

- Again right click on the *Coefficient Form PDE* tab in *Model Builder.* Select the *Dirichlet Boundary Condition* option. Another window will open namely *Dirichlet Boundary Condition 2.* Click on it.
- Select the bottom boundary on the square graph. Click on the "+" sign at the top right corner, this will add *boundary 2* in the *Boundary Selection* tab. At this point, put *r: T+Ty.* This step will add a *Neumann boundary condition* for the bottom boundary as given in Figure 5.12 (Note: as $T = r$, $r = T + Ty$, $T = Ty$).

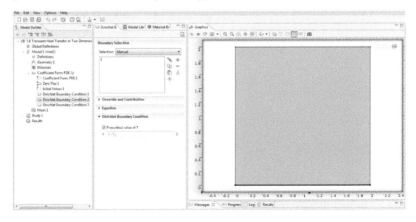

- Again right click on the *Coefficient Form PDE* tab in *Model Builder.* Select the *Dirichlet Boundary Condition* option. Another window will open namely *Dirichlet Boundary Condition 3.* Click on it.
- Select the top and bottom boundaries on the square graph. Click on the "+" sign at the top right corner, this will add *boundaries 3 and 4* in the *Boundary Selection* tab. At this point, put *r: 400.* This step will

add *Dirichlet boundary conditions* for the bottom and top boundaries as given in Figure 5.12. (Note: as $T = r$, $r = 400$, $T = 400$).

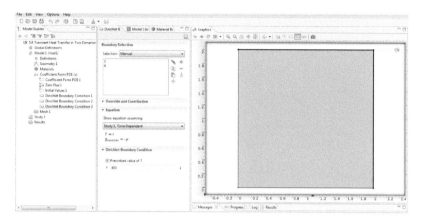

- Now, go to *Initial Values 1* available in the *Coefficient Form PDE* tab and set *T* to 0. This will add the initial condition given in the problem.

*Step 5*

- Click on the *Mesh* option in *Model Builder.* Select *Normal Mesh* Type. Click on *Build All* option at the top of ribbon. A dialogue box will appear in the *Message* tab as: *"Complete mesh consists of 578 elements."*

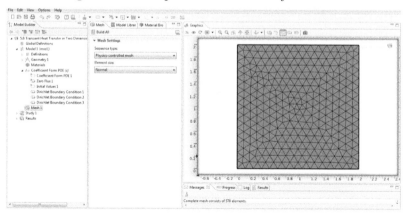

*Step 6*

- Click on the *Mesh* option in *Model Builder.* Select *Normal Mesh* Type. Click on *Build All* option at the top of ribbon. A dialogue box will appear in the *Message* tab as: *"Complete mesh consists of 578 elements."*

- Now, expand the *Study 1* tab in *Model Builder* and click on Step 1: Time Dependent and set the range (0,0.05,0.5) for Times. Click on the *Compute* (=) button. A graph will appear giving the profile of surface plot of *Temperature*.
- Save the simulation.

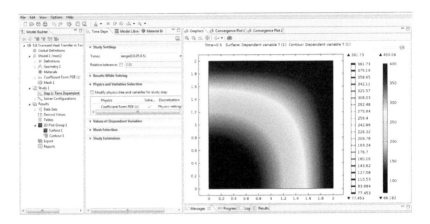

- Choose the *2D Plot Group* option by right clicking on *Results* tab. Another window will open namely *2D Plot Group 2*. Select *Contour* by right clicking on it. Another window will open namely *Contour 1* as shown in Figure 5.13.

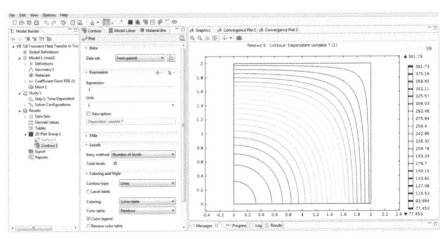

**FIGURE 5.13**   Solution to transient heat transfer problem.

## 5.9 PROBLEMS

1.  Determine the temperature in the following figure. The heat transfer is given by the equation

$$\frac{\partial^2 T}{\partial x^2} + \frac{\partial^2 T}{\partial y^2} = 0.$$

    The boundary conditions are shown below. The size of the slab is 2 m × 2 m, and step size in both the direction is $\Delta x = \Delta y = 0.5$ m.

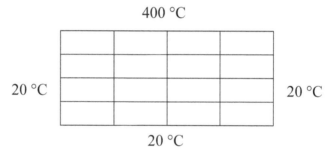

2.  Determine the temperature in the following figure. The heat transfer is given by the equation

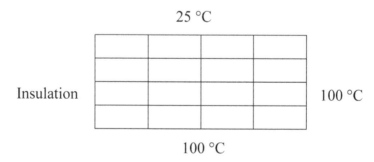

$$\frac{\partial^2 T}{\partial x^2} + \frac{\partial^2 T}{\partial y^2} = 0$$

    The boundary conditions are shown below. Take $\Delta x = \Delta y = 20$ cm.

3.  The slab is dried from both sides, is 4 cm thick, and has an initial uniform water concentration of 0.5 g of water per $cm^3$ (Carnahan et al., 1969). The drying will occur in the constant rate period at a rate of 0.1

g/cm² h of water as long as the surface moisture concentration remains above 0.2257 g of water per cm³. Take $\Delta x = \Delta y = 0.2$ cm, $\Delta t = 0.01$ h, and $D = 0.25$ cm²/h. Find the duration of the constant rate period and the distribution of water within the clay at the end. The partial differential equation is

$$\frac{\partial C}{\partial t} = D\left(\frac{\partial^2 C}{\partial x^2} + \frac{\partial^2 C}{\partial y^2}\right).$$

The boundary conditions are

$$\text{At } x = 0, \frac{\partial C}{\partial x} = 0.$$

$$\text{At } x = 2\,\text{cm}, -D\frac{\partial C}{\partial x} = 0.1.$$

$$\text{At } y = 0, \frac{\partial C}{\partial y} = 0.$$

$$\text{At } y = 2\,\text{cm}, -D\frac{\partial C}{\partial y} = 0.1.$$

4.  The diffusion equation for diffusion and reaction in a pore (Carnahan et al., 1969) is given by

$$\frac{d^2 C}{dx^2} + \frac{d^2 C}{dy^2} - \frac{k}{D}C = 0.$$

The boundary conditions are

$$\text{At } x = 0, C = C_s.$$

$$\text{At } x = L, \frac{dC}{dx} = 0.$$

$$\text{At } y = 0, C = C_s.$$

$$\text{At } y = L, \frac{dC}{dy} = 0.$$

$C_s$ is the concentration at the surface of the pore.
Pore length = 1 mm.
Rate constant of the reaction, $k = 10^{-3}$ s$^{-1}$
Effective diffusivity of species, $D = 10^{-9}$ m²/s.
Make 100 parts of the pore and determine the concentration at $x, y = 0.5$, 0.5 mm. The concentration at the surface of the mouth of the pore is 1 mol/m³.

5. Consider the convection and diffusion through a 2D space of a component of concentration $C$

$$-u\left(\frac{dC}{dx}+\frac{dC}{dy}\right)+D\left(\frac{d^2C}{dx^2}+\frac{d^2C}{dy^2}\right)=0.$$

The boundary conditions are

$$\text{At } x = 0, \, C = 1.$$
$$\text{At } x = 1 \text{ m}, \, C = 0.$$
$$\text{At } y = 0, \, C = 1.$$
$$\text{At } y = 1 \text{ m}, \, C = 0.$$

Calculate the distribution of $C$ as a function of $x$ *and* $y$. Make 10 parts between $x = 0$, $x = 1$ m and $y = 0$, $y = 1$ m.

6. Consider the differential equation for species $A$

$$D\left(\frac{d^2C_A}{dx^2}+\frac{d^2C_A}{dy^2}\right)-u\left(\frac{dC_A}{dx}+\frac{dC_A}{dy}\right)-kC_A=0$$

The boundary conditions are

$$\text{At } x = 0\,(\text{inlet}), uC_{A,\text{in}} = uC_A - D\frac{dC_A}{dx}.$$

$$\text{At } x = 10\,\text{m (exit)}, \frac{dC_A}{dx} = 0.$$

$$\text{At } y = 0\,(\text{inlet}), uC_{A,\text{in}} = uC_A - D\frac{dC_A}{dy}.$$

$$\text{At } y = 10\,\text{m (exit)}, \frac{dC_A}{dy} = 0.$$

A fluid medium comprising initially only $A$ flows the reactor with a mean axial velocity of $u = 1$ m/s. The axial dispersion coefficient, $D = 10^{-4}$ m$^2$ /s and the rate constant of the reaction is 0.1 s$^{-1}$. The inlet concentration $C_{A,\text{in}} = 1$ mol/m$^3$. Make 50, 20, and 10 parts of the reactor and determine the concentration of $A$ at various nodes along the tubular reactor of length 10 m.

7. A chemical reactor with axial dispersion is governed by the following governing equation:

$$\frac{1}{P_e}\left(\frac{d^2c}{dx^2}+\frac{d^2c}{dy^2}\right)-\frac{dc}{dx}-\frac{dc}{dy}-Da\frac{c}{c+v}=0.$$

The boundary conditions are

$$-\frac{1}{P_e}\frac{dc}{dx}(0) = 1 - c(0)$$

$$-\frac{1}{P_e}\frac{dc}{dy}(0) = 1 - c(0)$$

$$\frac{dc}{dx}(1) = 0$$

$$\frac{dc}{dy}(1) = 0.$$

Solve for $Da = 8$, $n = 3$, $P_e = 15$, $150$, $1500$.

8.  Consider reaction and diffusion in a packed bed with an initial concentration of zero in the bed. At t=0 start flowing with the inlet concentration =1.0. Use $P_e = 100$, $0 \le x \le 1$, $Da = 2$, $v = 2$.

$$\frac{\partial c}{\partial t} + \frac{dc}{dx} + \frac{dc}{dy} = \frac{1}{P_e}\left(\frac{d^2c}{dx^2} + \frac{d^2c}{dy^2}\right) - Da\frac{c}{c+v}.$$

The initial and boundary conditions are

$$c(x, 0) = 0$$

$$-\frac{1}{P_e}\frac{dc}{dx}(0,t) = 1 - c(0,t)$$

$$-\frac{1}{P_e}\frac{dc}{dy}(0,t) = 1 - c(0,t)$$

$$\frac{dc}{dx}(1,t) = 0$$

$$\frac{dc}{dy}(1,t) = 0.$$

# CHAPTER 6

# Optimization

## 6.1 INTRODUCTION

In this chapter, the application of *COMSOL* is discussed to solve optimization problems using *Optimization* module. The examples discussed here include optimal cooling of a tubular reactor, optimization of catalytic microreactor, and use of optimization in determination of Arrhenius parameters.

## 6.2 SIMULATION OF OPTIMAL COOLING OF A TUBULAR REACTOR

Figure 6.1 shows the tubular reactor in which two consecutive reactions take place. A heat exchanger jacket, run in co-current mode, is used to control the reaction rates and, hence, the product distribution in the reactor.

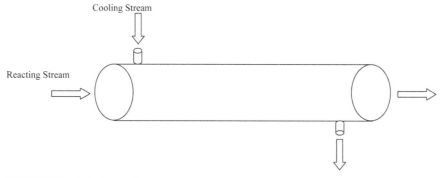

**FIGURE 6.1** Tubular reactor.

Temperature control in the reactor involves a delicate balance, where on the one hand, energy has to be supplied to the system to achieve acceptable reaction rates. On the other hand, the energy transfer to the reacting stream must be limited so that the desired intermediate product is not consumed by further reaction. The situation is further complicated by the fact that the temperature of the reacting stream is not only affected by the heat transfer from the heat exchanger jacket, but also by the endothermic nature of the reactions. The idea for this challenge in reactor optimization is taken from a literature, although the present reactor model is considerably more detailed.

The model is set up in 1D, coupling mass, and energy balances in the reactor tube with an energy balance for the heat exchanger jacket. Streams in both the tube and jacket are treated as plug flows.

### 6.2.1 CHEMISTRY

Two consecutive reactions occur in water, where the desired product is species $B$:

$$A \xrightarrow{k_1} B \tag{6.1}$$

$$B \xrightarrow{k_2} C. \tag{6.2}$$

The following rate equations apply:

$$r_1 = k_1 c_A \tag{6.3}$$

$$r_2 = k_2 c_B \tag{6.4}$$

where the rate constants are temperature dependent according to the Arrhenius relation:

$$k_j = A_j \exp\left(-\frac{E_j}{R_j T}\right) \tag{6.5}$$

$R_j = 8.314$ J/mol K.
The kinetic parameters are summarized in Table 6.1.

**TABLE 6.1**   Kinetic Parameters

| J | $A_j$ (1 s$^{-1}$) | $E_j$ (J/mol) |
|---|---|---|
| 1 | 1.6e8 | 75e3 |
| 2 | 1e15 | 125e3 |

### 6.2.2 MASS TRANSPORT

The convection–diffusion equation at steady state is used to model mass transport processes:

$$\nabla \cdot \left(-D_i \nabla c_i\right) + u \cdot \nabla c_i = R_i. \tag{6.6}$$

In this equation, $c_i$, $D_i$, *and* $R_i$ are concentration (mol/m³), diffusivity ($1 \times 10^{-8}$ m²/s) and rate expression for species $i$ (mol/(m³·s)), respectively. The velocity $u$ (m/s) of the fluid in the reactor is represented by a constant profile

$$u = 0.0042 \text{ m/s.} \tag{6.7}$$

At the inlet, the concentration of the reactant $A$ is 700 mol/m³. At the outlet, it is assumed that convective mass transport is dominant:

$$\nabla \cdot \left(-D_i \nabla c_i\right) = 0. \tag{6.8}$$

### 6.2.3  ENERGY TRANSPORT REACTOR

The energy transport in the reactor is described by

$$\nabla \cdot \left(-k \nabla T\right) + \rho C_p u \nabla T = Q_{rxn} + Q_j. \tag{6.9}$$

In the above equation, $k$, $T$, $\rho$, and $C_p$ are thermal conductivity (W/(m K)), temperature of the reacting stream (K), density (kg/m³), and heat capacity (J/(kg K)), respectively. The reacting species are diluted in water, and hence, the physical properties of the reacting mixture are assumed to be those of water.

The heat source due to reaction, $Q_{rxn}$ (W/m³), is calculated from the reaction rates and the enthalpies of reaction:

$$Q_{rxn} = \sum_{j=1-2} - \cdot H_j r_j. \tag{6.10}$$

Both of the reactions are endothermic, with $\Delta H_1 = 200$ kJ/mol and $\Delta H_2 = 100$ kJ/mol. Furthermore, the heat transferred from the reactor to the cooling jacket is given by

$$Q_j = -UA(T - T_j). \tag{6.11}$$

Here, $U$ is the overall heat transfer coefficient (J/(K m² s)), and $A$ represents the heat exchange area per unit volume (m²/m³), $UA = 10000$ W/m³ K.

The temperature of the reacting fluid at the inlet is 400 K. At the outlet, it is assumed that convective heat transport is dominant:

$$\nabla \cdot \left(-k \nabla T\right) = 0. \tag{6.12}$$

### 6.2.4  ENERGY TRANSPORT-COOLING JACKET

Water serves as the cooling medium in the jacket, and the energy transport is given by

$$\nabla \cdot \left(-k\nabla T_j\right) + \rho C_p u_j \nabla T_j = -Q_j. \tag{6.13}$$

The cooling stream is assumed to have plug flow character, and hence a constant velocity profile:

$$u_j = 0.01 \text{ m/s}. \tag{6.14}$$

The length of the reactor is 2 m. The optimal temperature of the cooling fluid at the inlet is to be found such that the maximum concentration of species $B$ is achieved at the outlet.

At the outlet, it is assumed that convective heat transport is dominant:

$$\nabla \cdot \left(-k\nabla T\right) = 0. \tag{6.15}$$

### 6.2.5  SIMULATION APPROACH

*Step 1*

- Open COMSOL Multiphysics.
- Select *1D Space Dimension* from the list of options. Hit the *next* arrow at the upper right corner.

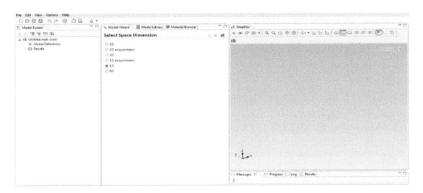

- Expand the *Chemical Species Transport* folder from the list of options in *Model Wizard* and select *Transport of Diluted Species (chds)*. Click on add selected button. In the Dependent variables tab, change *Number of species* to 3. Change the species concentration to "$c_A$, $c_B$, and $c_C$."

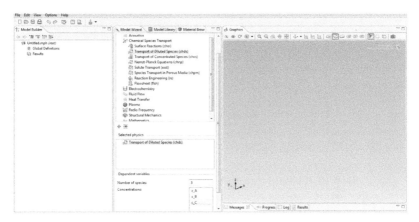

- In the *Add Physics* tab, expand the *Heat Transfer* folder from the list of options in *Model Wizard* and select *Heat Transfer in Fluids*. Click on add selected button. In the Dependent variables tab, keep the Temperature to "$T$."

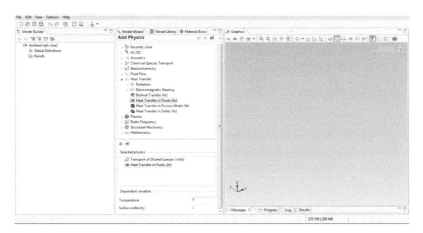

- Again in the *Add Physics* tab, expand the *Heat Transfer* folder from the list of options in *Model Wizard* and select *Heat Transfer in Fluids*. Click on add selected button. In the Dependent variables tab, change the Temperature to "$T_j$".

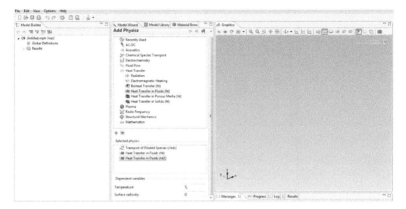

- Again in the *Add Physics* tab, expand the *Mathematics* folder from the list of options in *Model Wizard* and select *Optimization*. Click on add selected button.

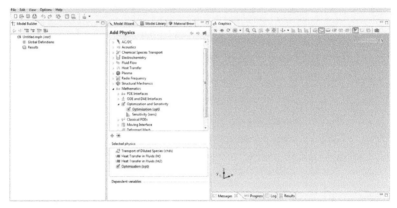

- Hit the *next* arrow and select *Stationary* from the list of *Study Type* options, and click on the *Finish flag* at the upper right corner of the application.

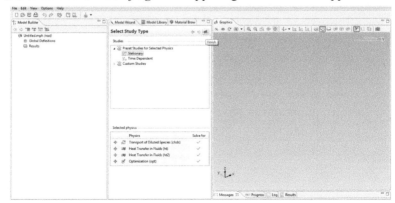

*Step 2*

- With the *Model Builder*, right click on the *Geometry* option. From the list of options, click on the *Interval* taskbar.
- In the *Interval* environment, select *Intervals1, Left endpoint: 0, and Right endpoint: 2*. Click on the *Build All* option at the upper part of tool bar. A straight line graph will appear on the right side of the application window.

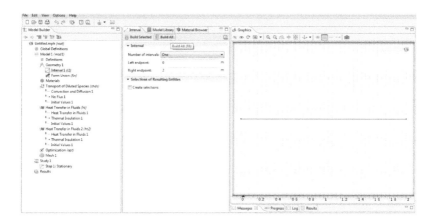

*Step 3*

- Now in the *Model Builder*, select *Convection and Diffusion 1* option available in the *Transport of Diluted Species* option. This will open a tab to enter the coefficients of the characteristics of a model equation.
- In the *Domain Selection* panel, you will see an equation of the form

$$\nabla \cdot \left(-D_i \nabla c_i\right) + u \cdot \nabla c_i = R_i \tag{6.16}$$

$$N_i = -D_i \nabla c_i + u c_i.$$

This is same as Equation (6.6). In order to solve the governing differential equation, we need to assign the coefficients in above equation a suitable value.

Set u as user defined and diffusion coefficient $Dc_A$, $Dc_B$, $Dc_C$ to user defined value $D$.

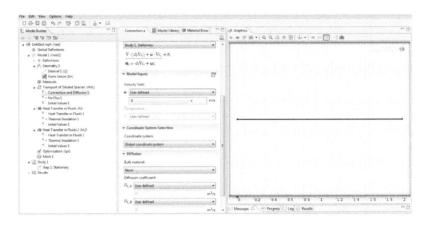

- With the *Model Builder*, right click on the *Global Definitions* tab and choose *Parameters*. Another window will open namely *Parameters*. Click on it and define *D: 1e-8, u: 0.0042, u_j: 0.01, T_{in}: 400, UA: 10,000, H1: 200e3, H2: 100e3, Rg: 8.314, A1: 1.6e8, A2: 1e15, E1: 75e3, and E2: 125e3.*

- With the *Model Builder*, right click on the *Definitions* tab and choose *Variables*. Another window will open namely *Variables 1*. Click on it and define $k1$: $A1*\exp(-E1/(Rg*T))$, $k2$: $A2*\exp(-E2/(Rg*T))$, $r1$: $k1*c^A$, $r2$: $k2*c^B$, and $Q_{rxn}$: $-r1*H1 - r2*H2$.

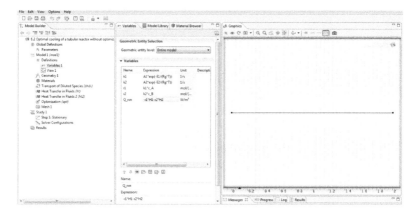

- With the *Model Builder*, right click on the *Materials* tab and choose *Open Material Browser*. Another window will open namely *Materials*. In the *Materials* section, select Liquids and Gases > Liquids > Water.

- Right click on it and select *Add Material*.

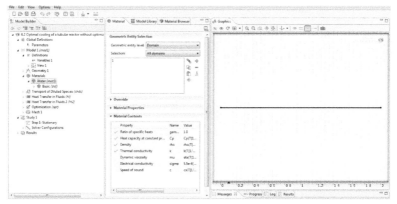

*Step 4*

- Now, right click on the *Transport of Diluted Species (chds)* tab. Choose the *Inflow* option. Another window will open namely *Inflow 1*. Click on it.
- Select the left point on the horizontal line graph. Click on the "+" sign at the top right corner, this will add *boundary 1* in the *Boundary Selection* tab. At this point, put $c_{0,c\_A}$: 700, $c_{0,c\_B}$: 0, $c_{0,c\_C}$: 0. This step will add an inlet boundary condition of Equation (6.6).

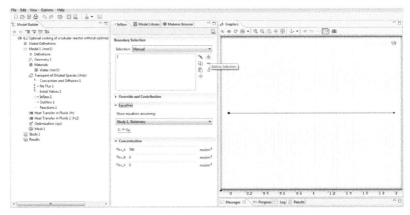

- Again right click on the *Transport of Diluted Species (chds)* tab. Choose the *Outflow* option. Another window will open namely *Outflow 1*. Click on it.
- Select the right point on the horizontal line graph. Click on the "+" sign at the top right corner, this will add *boundary 2* in the *Boundary Selection* tab. This will add the equation $-n.D_i\nabla_{ci} = 0$, which is same as the outlet boundary condition as given in Equation (6.8).

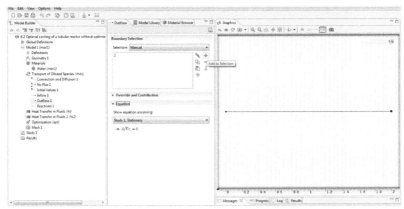

- Again right click on the *Transport of Diluted Species (chds)* tab. Choose the *Reactions* option. Another window will open namely *Reaction 1*. Click on it.
- Select *Reactions1* option available in the *Transport of Diluted Species* tab and expand the Reactions tab. Change the $R_{c\_A}$: $-r1$, $R_{c\_B}$: $r1 - r2$, $R_{c\_C}$: $r2$. Select the horizontal line graph. Click on the "+" sign at the top right corner, this will add a horizontal line in the *Domain Selection* tab.

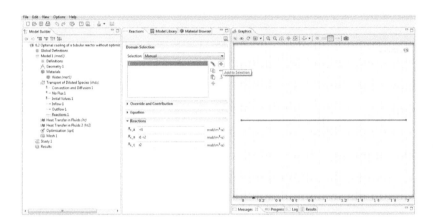

## Step 5

- Now, in the *Model Builder*, select *Heat Transfer in Fluids 1* option. This will open a tab to enter the coefficients of the characteristics of a model equation.
- In the *Domain Selection* panel, you will see an equation of the form

$$\rho C_p u \cdot \nabla T = \nabla \cdot (k \nabla T)) + Q + Q_{vh} + W_p. \tag{6.17}$$

- In order to solve the governing differential equation, we need to assign the coefficients in above equation a suitable value.
- To convert Equation (6.17) to the desired form of Equation (6.9), the value of coefficients in Equation (6.17) to be changed as follows:
- Set $u$ as "User defined". Set $k$, $\rho$, $C_p$. and $\gamma$ as "From material."

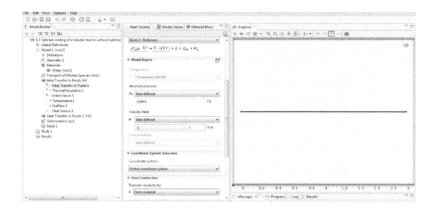

*Step 6*

- Now, right click on the *Heat Transfer in Fluids* tab in *Model Builder.* Select the *Temperature* option. Another window will open namely *Temperature 1.* Click on it.
- Select the left point on the horizontal line graph. Click on the "+" sign at the top right corner, this will add *boundary 1* in the *Boundary Selection* tab. At this point, put $T_0$: $T_{in}$. This step will add an inlet boundary condition of Equation (6.9).

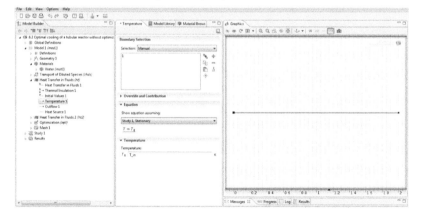

- Again right click on the *Heat Transfer in Fluids* tab in *Model Builder.* Select the *Outflow* option. Another window will open namely *Outflow 1.* Click on it.
- Select the right point on the horizontal line graph. Click on the "+" sign at the top right corner, this will add *boundary 2* in the *Boundary Selection* tab. This will add the equation $-n.(k\nabla_T) = 0$, which is same as the outlet boundary condition as given in Equation (6.12).

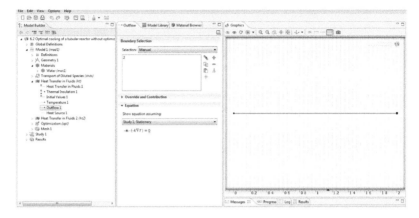

- Again right click on the *Heat Transfer in Fluids* tab in *Model Builder.* Select the *Heat Source* option. Another window will open namely *Heat Source 1.* Click on it.
- Select the horizontal line graph. Click on the "+" sign at the top right corner, this will add a horizontal line in the *Domain Selection* tab. Set the value of the Heat Source $Q: -UA*(T - T_j)+Q_{rxn}.$

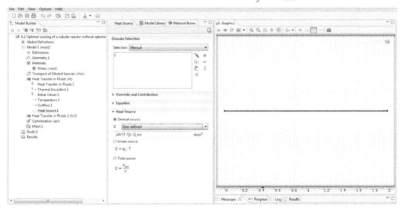

*Step 7*

- Now in the *Model Builder,* select *Heat Transfer in Fluids 1* option available in *Heat Transfer in Fluids 2.* This will open a tab to enter the coefficients of the characteristics of a model equation.
- In the *Domain Selection* panel, you will see an equation of the form

$$\rho C_p u \cdot \nabla Tj = \nabla \cdot (k \nabla Tj) + Q + Q_{vh} + W_p. \tag{6.18}$$

In order to solve the governing differential equation, we need to assign the coefficients in above equation a suitable value.

To convert Equation (6.18) to the desired form of Equation (6.13), the value of coefficients in Equation (6.18) to be changed as follows: Set $u$ as "User defined" $u_j$. Set $k$, $\rho$, $C_p$, and $\gamma$ as "From material."

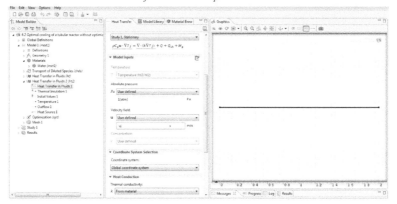

*Step 8*

- Now, right click on the *Heat Transfer in Fluids 2* tab in *Model Builder.* Select the *Temperature* option. Another window will open namely *Temperature 1.* Click on it.
- Select the left point on the horizontal line graph. Click on the "+" sign at the top right corner, this will add *boundary 1* in the *Boundary Selection* tab. At this point, put $T_0$: $Tj_{in}$. The optimal temperature of the cooling fluid $T_j$ at the inlet is to be found such that the maximum concentration of species $B$ is achieved at the outlet.

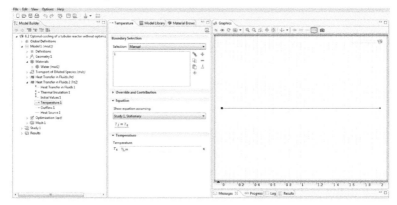

- Again right click on the *Heat Transfer in Fluids 2* tab in *Model Builder.* Select the *Outflow* option. Another window will open namely *Outflow 1.* Click on it.

- Select the right point on the horizontal line graph. Click on the "+" sign at the top right corner, this will add *boundary 2* in the *Boundary Selection* tab. This will add the equation $- n.(k \nabla_T) = 0$, which is same as the outlet boundary condition as given in Equation (6.15).

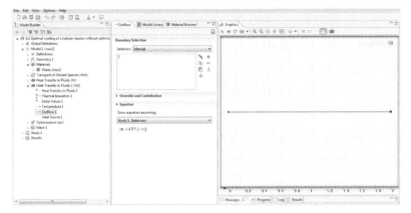

- Again right click on the *Heat Transfer in Fluids 2* tab in *Model Builder*. Select the *Heat Source* option. Another window will open namely *Heat Source 1*. Click on it.
- Select the horizontal line graph. Click on the "+" sign at the top right corner, this will add a horizontal line in the *Domain Selection* tab. Set the value of the Heat Source $Q: UA*(T - T_j)$.

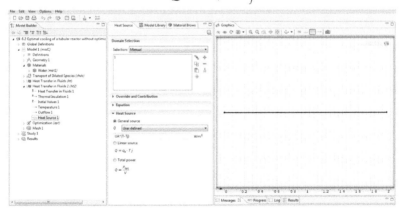

*Step 9*

- Now, right click on the *Optimization* tab in *Model Builder*. Select the *Global Control Variables* option. Another window will open namely *Global Control Variables 1*. At this point, put $Tj_{in}: 400$.

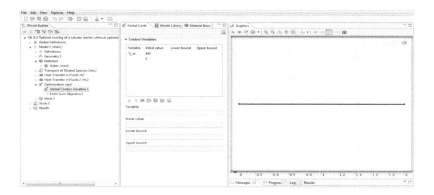

- Again right click on the *Optimization* tab in *Model Builder.* Select the *Point Sum Objective* option. Another window will open namely *Point Sum Objective 1.*
- Select the right point on the horizontal line graph. Click on the "+" sign at the top right corner, this will add *boundary 2* in the *Boundary Selection* tab. At this point, put *Objective expression:* $-c_B$.

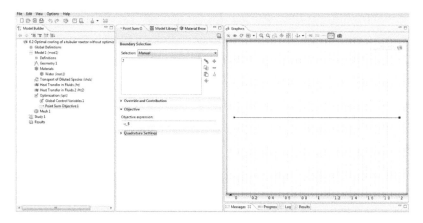

*Step 10*

- With the *Model Builder*, right click on the *Mesh 1* tab and choose the *Edge* option.
- With the *Model Builder*, again right click on the *Mesh 1* tab and choose the *Size* option. Select the *Element Size* as *Extrafine.* Click on *Build All* option at the top of ribbon. A dialogue box will appear in the *Message* tab as: *"Complete mesh consists of 50 elements."*

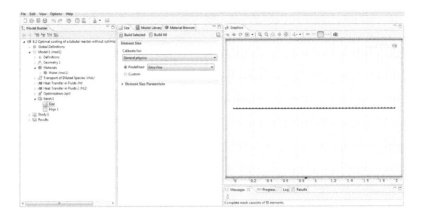

## Step 11

- Now, go to *Study 1* option in the model pellet tab. Click on the *Compute (=)* button. To plot $c_A$ versus $x$, right click on the *Concentration* and select the *Line Graph*. Another *Plot* window with the name *Line Graph 1* will open. Change the *Expression* in the *y-Axis Data* to $c_A$. Now go to *Study* option in the model pellet tab. Click on the *Plot* button. A graph will appear giving the profile of $c_A$ versus $x$.
- Repeat the above step to plot $c_B$ versus $x$ and $c_C$ versus $x$.
- Save the simulation.

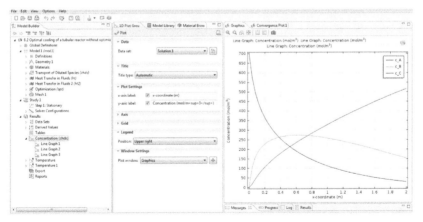

**FIGURE 6.2** Species concentrations (blue $c_A$, green $c_B$, red $c_C$) as function of reactor position when the inlet temperature of the cooling fluid is 400 K.

- To plot $T$ versus $x$, right click on the *Temperature* and select the *Line Graph*. Another *Plot* window with the name *Line Graph 1* will open. Change the *Expression* in the *y-Axis Data* to $T$. Now, go to *Study*

option in the model pellet tab. Click on the *Plot* button. A graph will appear giving the profile of *T versus x.*
- Repeat the above step to plot $T_j$ *versus x.*
- Save the simulation.

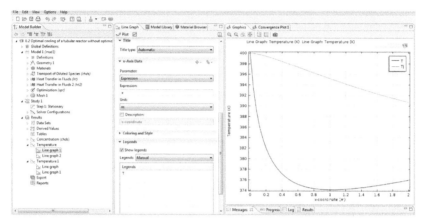

**FIGURE 6.3**  Temperature distribution for the reacting stream (blue) and jacket stream (green) when the inlet temperature of the jacket stream is 400 K.

- To plot reaction rate ($r_1$) *versus* x, right click on the *Temperature 1* and select the *Line Graph.* Another *Plot* window with the name *Line Graph 1* will open. Change the *Expression* in the *y-Axis Data* to $r_1$. Now go to *Study* option in the model pellet tab. Click on the *Plot* button. A graph will appear giving the profile of $r_1$ versus *x.*
- Repeat the above step to plot $r_2$ versus *x.*
- Save the simulation.

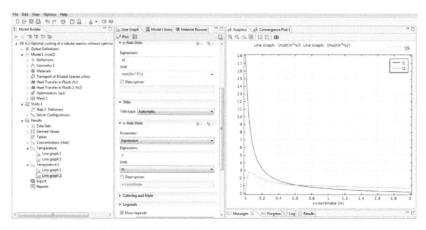

**FIGURE 6.4**  Rate of the production $r_1$ (blue) and rate consumption $r_2$ (green) of species *B* when the inlet temperature of the cooling fluid is 400 K.

*Step 12*

- Now, solve the optimization problem.
- Go to *Stationary* tab under *Study 1* in *Model Builder.* Expand the *Study Extensions* and select *Optimization* check box.

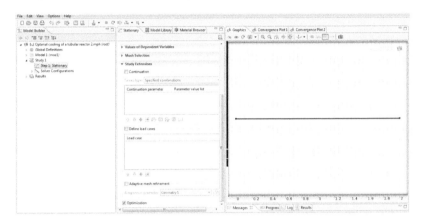

- With the *Model Builder*, right click on *Study 1* and choose *Show Default Solver.* In the *General* section of *Optimization Solver 1* node, change the *Optimality tolerance* to 0.01.

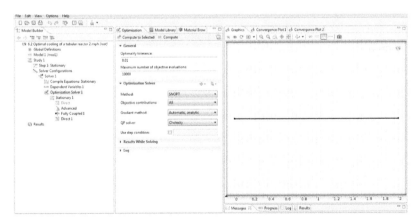

- With the *Model Builder*, expand the *Stationary 1* node under *Optimization Solver 1* and choose *Fully Coupled 1*. In the *Fully Coupled 1* window, expand the *Method and Termination* section and choose *Constant (Newton)*.

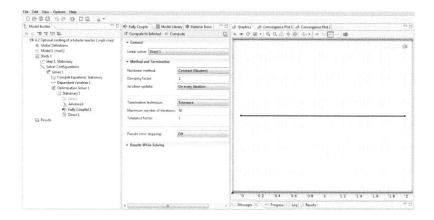

- Now, go to *Study 1* option in the model pellet tab. Click on the *Compute (=)* button.

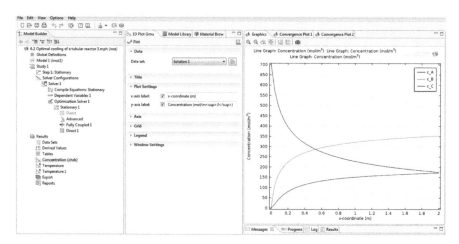

**FIGURE 6.5**   Species concentrations (blue $c_A$, green $c_B$, red $c_C$) as function of reactor position when the inlet temperature of the cooling fluid is 334 K.

- In a first simulation, the inlet temperatures of the jacket stream and the reacting stream are set to be equal, at 400 K. In a second simulation, an optimization calculation is performed to find the inlet temperature of the jacket stream that maximizes the concentration of the desired intermediary product (*B*) at the reactor outlet. Comparisons between the two cases follow below.

- Figure 6.2 shows the concentration of reacting species as a function of the reactor length when the inlet temperature of the jacket stream is 400 K. Figure 6.3 shows concentration curves for the optimal inlet temperature of the jacket stream, found to be 334 K. Clearly, when the inlet temperature is 400 K the conversion of reactant $A$ is high, but at the same time, the selectivity for the desired product $B$ is unfavorable. Under the optimized conditions, the concentration of $B$ at the reactor outlet is 353.5 mol/m³, to be compared to a concentration of 153.8 mol/m³ when the inlet temperature is 400 K.
- Plots of reacting stream and jacket stream temperatures are shown in Figures 6.4 and 6.5. The jacket stream heats up the reacting stream when its inlet temperature is kept at 400 K.
- In contrast, the jacket stream cools the reacting stream when its inlet temperature is 334 K.

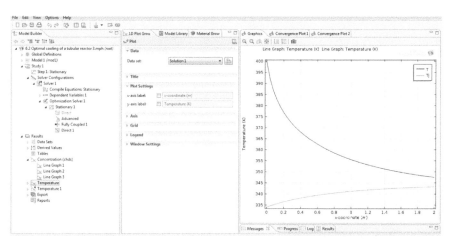

**FIGURE 6.6**   Temperature distribution for the reacting stream (blue) and jacket stream (green) when the inlet temperature of the jacket stream is 334 K.

- The reaction rates are illustrated in Figures 6.4 and 6.7. When the inlet temperature of the jacket stream is 400 K, the rate at which $B$ is consumed $(r_2)$ dominates over the production rate $(r_1)$ from a point approximately 0.65 m down the reactor. This effect is due to heat being transferred from the jacket stream, counteracting the cooling effect of the endothermic reactions.

- At an inlet temperature of 334 K, the combined effect of cooling by the jacket stream and energy consumption due to reaction work together to quench the system, resulting in increased concentrations levels of $B$ at the outlet.

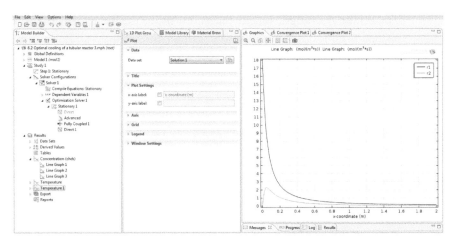

**FIGURE 6.7**   Rate of the production $r_1$ (blue) and rate consumption $r_2$ (green) of species $B$ when the inlet temperature of the cooling fluid is 334 K.

## 6.3   SIMULATION OF OPTIMIZATION OF A CATALYTIC MICROREACTOR

### 6.3.1   PROBLEM STATEMENT

Consider that the reaction is carried out when the solution is pumped through the catalyst bed. The reaction rate has to be maximized by considering the optimal distribution of catalyst across the bed. The low flow rate is resulted if large amount of catalyst is used and vice versa.

*Model Definition*

The model geometry is depicted in Figure 6.8. The reactor consists of an inlet channel, a fixed catalytic bed, and an outlet channel.

It is represented in COMSOL by three rectangles with the following dimensions:

Rectangle 1: Width = $2*L$, Height = $L$, Base corner: at $x = 0$, $y = 0$.
Rectangle 2: Width = $6*L$, Height = $3*L$, Base corner: at $x = 2*L$, $y = 0$.
Rectangle 3: Width = $2*L$, Height = $L$, Base corner: at $x = 8*L$, $y = 0$.

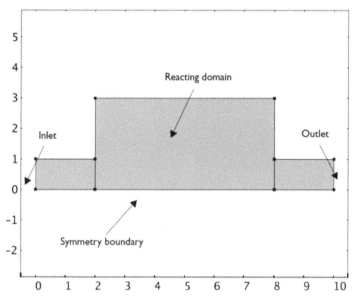

**FIGURE 6.8**   The model geometry.

The objective is to maximize the average reaction rate by optimizing the catalyst distribution. The average reaction rate is expressed as the integral of the local reaction rate, $r$ (mol/(m$^3$·s)), over the domain, $\Omega$. It is to be maximized or minimize the negative of it

$$\min\left\{-\frac{1}{\mathrm{vol}(\Omega)}\int_{\Omega}rd\Omega\right\}. \tag{6.19}$$

The local reaction rate is determined by

$$r = k_a(1-\varepsilon)c. \tag{6.20}$$

In this equation, $\varepsilon$, $c$, and $k_a$ are the volume fraction of solid catalyst, concentration (mol/m$^3$) and rate constant (0.25 1/s), respectively.

The mass transport is described by the convection and diffusion equation

$$\nabla\cdot\left(-D\nabla c\right) = r - u\nabla c. \tag{6.21}$$

In this equation, $u$ and $D$ are velocity vector (m/s) and diffusion coefficient ($3\times10^{-8}$ m$^2$/s).

Inlet from the left boundary:

$$\text{At } x = 0, c_{in} = 1\frac{\text{mol}}{\text{m}^3}\text{ at any } y. \tag{6.22a}$$

At the outlet from the right boundary, it is assumed that convective mass transport is dominant:

$$\nabla \cdot \left( -D_i \nabla c_i \right) = 0. \tag{6.22b}$$

The Navier–Stokes equations for 2D, laminar, incompressible flow with constant viscosity are described by

$$\rho\left(u \cdot \nabla\right)u = -\nabla p + \nabla \cdot \eta \left( \nabla u + \left( \nabla u \right)^T \right) - \alpha\left(\varepsilon\right)u \tag{6.23}$$

$$\nabla u = 0. \tag{6.24}$$

The above equations can be rewritten as

$$\rho\left( u\frac{\partial u}{\partial x} + v\frac{\partial u}{\partial y} \right) = -\frac{\partial p}{\partial x} + \eta\left( \frac{\partial^2 u}{\partial x^2} + \frac{\partial^2 u}{\partial y^2} \right) - \alpha\left(\varepsilon\right)u \tag{6.25}$$

$$\frac{\partial u}{\partial x} + \frac{\partial v}{\partial y} = 0. \tag{6.26}$$

The coefficient α(ε) depends on the distribution of the porous catalyst as

$$\alpha\left(\varepsilon\right) = \frac{\eta}{D_a L^2}\frac{q(1-\varepsilon)}{q+\varepsilon}. \tag{6.27}$$

Boundary conditions:
Inflow from the left:

$$\text{At } x = 0,\ p_0 = 0.25 \text{ Pa at any } y. \tag{6.28a}$$

Outflow from the right:
At the outflow, the viscous stress in fully developed flow is considered; hence, pressure is zero.

$$\text{At } x = 10 * L,\ p_0 = 0.25 \text{ Pa at any } y. \tag{6.28b}$$

Inflow from the bottom:

$$\text{At } y = 0,\ u = v = 0 \text{ at any } x \tag{6.28c}$$

The wall at all other boundaries is having no slip condition with the velocity zero for both components.

Where $D_a$ is the Darcy number ($1 \times 10^{-4}$), $L$ is the length scale (1 mm), $q$ is a dimensionless optimization parameter (0.04), the interpretation of which is discussed in the next section; $\rho$ is density of water (1000 kg/m$^3$), eta is viscosity of water ($1 \times 10^{-3}$ Pa s), and $\nabla p$ is pressure drop (0.25 Pa).

From Equation (6.27), the direct conclusion is that when ε equals 1, α equals zero, and Equation (6.25) reduces to the ordinary Navier–Stokes equations. In this case, the reaction rate is zero; see Equation (6.20).

To summarize, the optimization problem is

$$\min\left\{-\frac{1}{\text{vol}(\Omega)}\int_\Omega k_a(1-\varepsilon)cd\Omega\right\}$$

where

$$\rho(u\cdot\nabla)u = -\nabla p + \nabla\cdot\eta\left(\nabla u + (\nabla u)^T\right) - \alpha(\varepsilon)u$$

$$\nabla u = 0$$

$$\nabla\cdot(-D\nabla c) = r - u\nabla c$$

$$0 \le \varepsilon \le 1$$

and physical boundary conditions apply. Vol is volume of reaction domain ($3*L*6*L$).

### 6.3.2 CONVEX OPTIMIZATION PROBLEMS

One of the most important characteristics of an optimization problem is whether or not the problem is convex. This section therefore briefly describes this property.

A set $C$ is said to be convex if for any two members $x$ and $y$ of $C$, the following relation holds:

$$tx + (1-t)y \in C \text{ for every } t \in [0,1]$$

that is, the straight line between $x$ and $y$ is fully contained in $C$. A convex function is a mapping $f$ from a convex set $C$ such that for every two members $x$ and $y$ of $C$

$$f(tx+(1-t)y) \le tf(x)+(1-t)f(y) \text{ for every } t \in [0,1]. \quad (6.29)$$

An optimization problem is said to be convex if the following conditions are met:

- Design domain is convex.
- The objective and constraints are convex functions.

The importance of convexity follows simply from the result that if $x^*$ is a local minimum to a convex optimization problem, then $x^*$ is also a global minimum. This is easily proven by simply assuming that there is a $y$ such that $f(y) < f(x^*)$, and then using Equation (6.29).

This particular optimization problem is nonlinear, because a change in $\varepsilon$ implies a change in the concentration, $c$. Because of this implicit dependence, it is very difficult to determine whether or not the objective is convex. There is therefore no guarantee that the optimal solution you obtain is globally optimal or unique. In the best of cases, running the optimization will give a good local optimum.

The parameter q can be used to smoothen the interfaces between the catalyst and the open channel.

### 6.3.3   *SIMULATION APPROACH*

*Step 1*

- Open COMSOL Multiphysics.
- Select *2D Space Dimension* from the list of options. Hit the *next* arrow at the upper right corner.

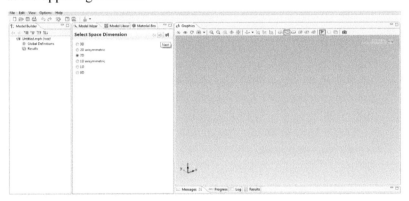

- Expand the *Fluid Flow* folder from the list of options in *Model Wizard* and select *Laminar Flow* available under *Single-Phase Flow*. Click on add selected button.

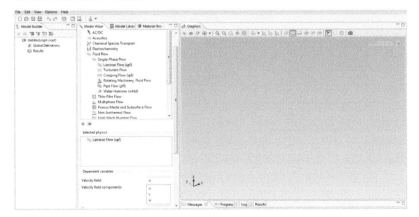

- Again in the *Add Physics* tab, expand the *Chemical Species Transport* folder from the list of options in *Model Wizard* and choose *Transport of Diluted Species (chds)*. Click on add selected button.

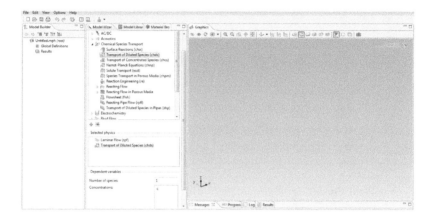

- Again in the *Add Physics* tab, expand the *Mathematics* folder from the list of options in *Model Wizard* and select *Optimization*. Click on add selected button.

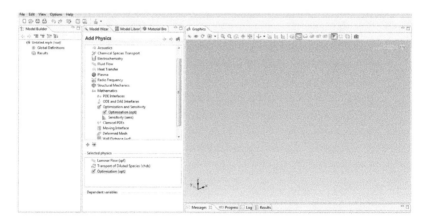

- Hit the *next* arrow and select *Stationary* from the list of *Study Type* options, and click on the *Finish flag*.

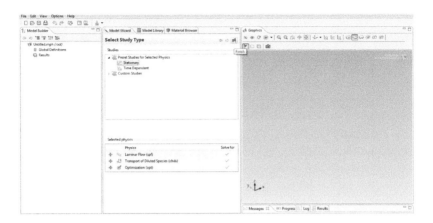

*Step 2*

- With the *Model Builder*, right click on the *Global Definitions* tab and select *Parameters*. Another window will open namely *Parameters*. Click on it and define:

$k_a$ is the rate constant (0.25 1/s)

$D$ is the diffusion coefficient ($3 \times 10^{-8}$ m²/s)

$D_a$ is the Darcy number ($1 \times 10^{-4}$)

c_in is the concentration at the inlet (1 mol/m³)

$L$ is the length scale (1 mm)

$q$ is a dimensionless optimization parameter (0.04)

*rho* is density of water (1000 kg/m³)

eta is viscosity of water ($110^{-3}$ Pa.s)

delta$_p$ is pressure drop (0.25 Pa)

vol is volume of reaction domain ($3*L*6*L$).

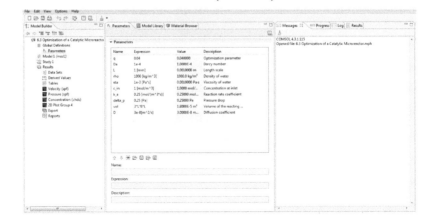

*Step 3*

- With the *Model Builder*, click on the *Geometry 1* option. In the *Geometry 1* setting window, change the *Length unit* to mm.
- With the *Model Builder*, again right click on the *Geometry 1* option. From the list of options, click on the *Rectangle* taskbar.
- In the *Rectangle 1* environment, select *Size* with *Width: 2\*L, Height: L, Base Corner: x = 0, y = 0*. Click on the *Build All* option at the upper part of tool bar. A rectangle graph will appear on the right side of the application window.

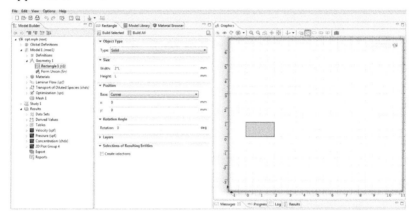

- With the *Model Builde*, again right click on the *Geometry* option. From the list of options, click on the *Rectangle* taskbar.
- In the *Rectangle 2* environment, select *Size* with *Width: 6\*L, Height: 3\*L, Base Corner: x = 2\*L, y = 0*. Click on the *Build All* option at the upper part of tool bar. A rectangle graph will appear on the right side of the application window.

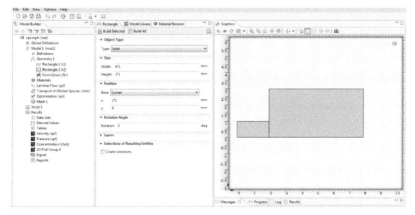

- With the *Model Builder*, again right click on the *Geometry* option. From the list of options, click on the *Rectangle* taskbar.
- In the *Rectangle 3* environment, select *Size* with *Width: 2\*L, Height: L, Base Corner: x = 8\*L, y = 0*. Click on the *Build All* option at the upper part of tool bar. A rectangle graph will appear on the right side of the application window.

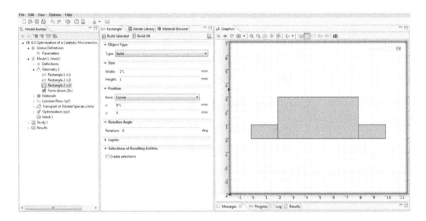

- Now, we define integration couplings to use for calculating the conversion of the reactant.
- With the *Model Builder*, right click on the *Definitions* tab and select *Model Couplings>Integration*. Another window will open namely *Integration 1*. Select *Boundary* from the *Geometric entity level*.
- Select the left boundary on the graph. Click on the "+" sign at the top right corner, this will add *boundary 1* in the *Boundary Selection* tab.

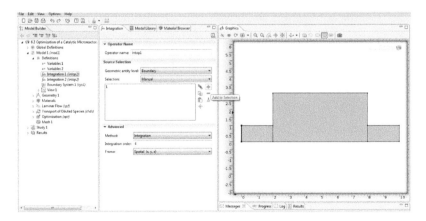

- With the *Model Builder*, again right click on the *Definitions* tab and select *Model Couplings > Integration*. Another window will open namely *Integration 2*. Select *Boundary* from the *Geometric entity level*.
- Select the right boundary on the graph. Click on the "+" sign at the top right corner, this will add *boundary 12* in the *Boundary Selection* tab.

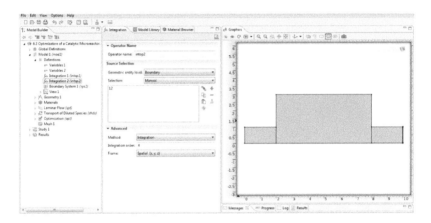

- With the *Model Builder*, again right click on the *Definitions* tab and select *Variables*. Another window will open namely *Variables 1*. Click on it and define $F_{in}$: *intop1(chds.tfluxx_c)*, $F_{out}$: *intop2(chds.tfluxx_c)*, X: $(F_{in} - F_{out})/F_{in}$.
- Here, chds.tfluxx$_c$ is the COMSOL Multiphysics variable for the $x$ component of the total flux.

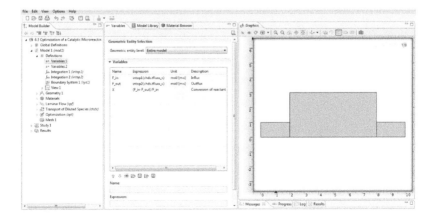

- With the *Model Builder*, again right click on the *Definitions* tab and select *Variables*. Another window will open namely *Variables 2*. Click on it and define *phi: $k_a*(1 – epsilon)*c$, alpha: $(eta/(Da*L^2))*q*(1 – epsilon)/(q + epsilon)$*. Select the *Domain 2* from the graph. Click on the "+" sign at the top right corner, this will add *Domain 2* in the *Domain Selection* tab.

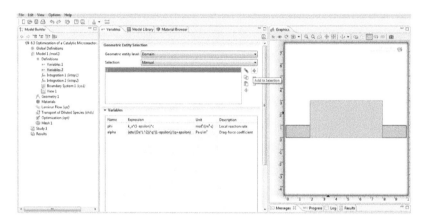

- With the *Model Builder*, again right click on the *Materials* tab and select *Open Material Browser*. Another window will open namely *Materials*. In the *Materials* section, select Liquids and Gases > Liquids > Water.

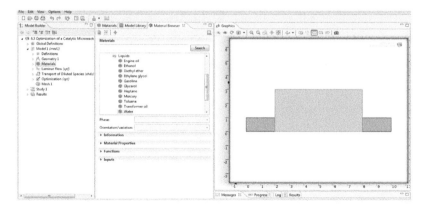

- Right click on it and choose *Add Material*.

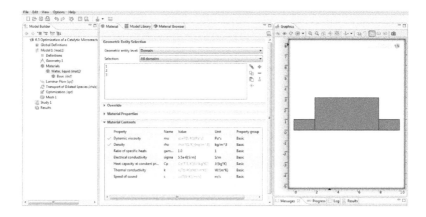

## Step 4

- Now, with the *Model Builder*, select *Fluid Properties 1* option available in *Laminar Flow*. This will open a tab to enter the coefficients of the characteristics of a model equation.
- In the *Domain Selection* panel, you will see an equation of the form

$$\rho(u \cdot \nabla)u = \nabla \cdot \left[ -pl + \mu\left( \nabla u + (\nabla u)^T - \frac{2}{3}\mu(\nabla \cdot u)l \right) \right] + F \qquad (6.30)$$

$$\nabla \cdot (\rho u) = 0. \qquad (6.31)$$

In order to solve the governing differential equation, we need to assign the coefficients in above equation a suitable value.

To convert Equations (6.30) and (6.31) to the desired form of Equations (6.25) and (6.26), respectively, the value of coefficients in Equations (6.30) and (6.31) to be changed as follows:

*Take $\rho$ and $\mu$ from the material.*

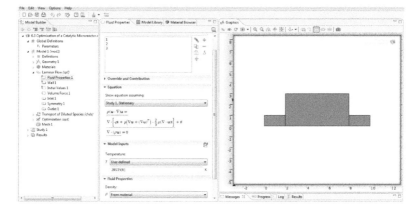

- Now, right click on the *Laminar Flow* tab in *Model Builder.* Select the *Volume Force* option. Another window will open namely *Volume Force 1.* Click on it.
- Select the *Domain 2* from the graph. Click on the "+" sign at the top right corner, this will add *Domain 2* in the *Domain Selection* tab. At this point, put *Volume Force F: –alpha\*u for x and –alpha\*v for y.*

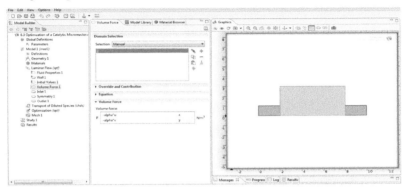

## Step 5

- Now, right click on the *Laminar Flow* tab in *Model Builder.* Select the *Inlet* option. Another window will open namely *Inlet 1.* Click on it.
- Select the left boundary on the graph. Click on the "+" sign at the top right corner, this will add *boundary 1* in the *Boundary Selection* tab. At this point, put *Pressure $p_0$: delta$_p$.* This step will add a boundary condition as given in Equation (6.28a).

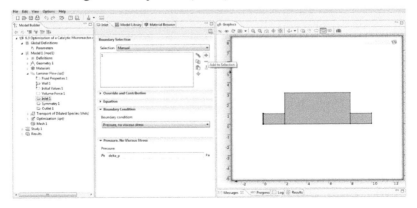

- Again right click on the *Laminar Flow* tab in *Model Builder.* Select the *Outlet* option. Another window will open namely *Outlet 1.* Click on it.

- Select the right boundary on the graph. Click on the "+" sign at the top right corner, this will add *boundary 12* in the *Boundary Selection* tab. At this point, put *Pressure $p_0$: 0.* This step will add a boundary condition as given in Equation (6.28b).

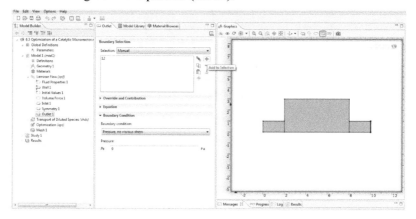

- Again right click on the *Laminar Flow* tab in *Model Builder.* Select the *Symmetry* option. Another window will open namely *Symmetry 1.* Click on it.
- Select the all the bottom boundaries on the graph. Click on the "+" sign at the top right corner, this will add *boundaries 2, 5, and 9* in the *Boundary Selection* tab. This step will add a boundary condition as given in Equation (6.28c).

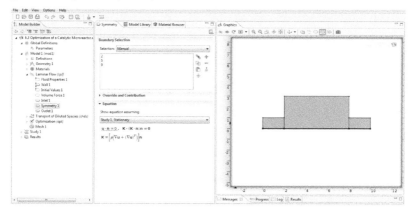

- Now, in the *Model Builder*, select *Wall 1* option available in *Laminar Flow*. By default it has all other boundaries in the *Boundary Selection* tab. At this point, put boundary condition to *No slip*.

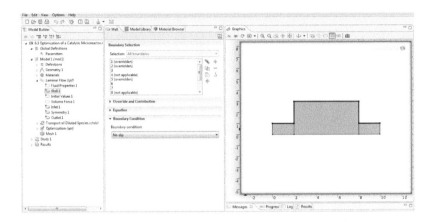

*Step 6*

- Now, in the *Model Builder*, select *Convection and Diffusion 1* option available in the *Transport of Diluted Species* option. This will open a tab to enter the coefficients of the characteristics of a model equation.
- In the *Domain Selection* panel, you will see an equation of the form

$$\nabla \cdot \left( -D_i \nabla c_i \right) + u \cdot \nabla c_i = R_i \qquad (6.32)$$

$$N_i = -D_i \nabla c_i + u c_i$$

This is same as Equation (6.21). In order to solve the governing differential equation, we need to assign the coefficients in above equation a suitable value.

Choose *Velocity field* from **u** list and set diffusion coefficient $D_c$ to user defined value $D$.

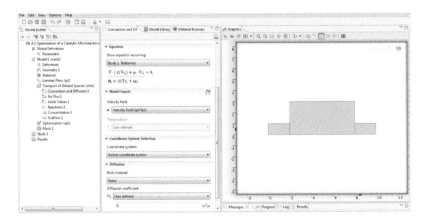

*Step 7*

- Now, right click on the *Transport of Diluted Species (chds)* tab in *Model Builder.* Select the *Concentration* option. Another window will open namely *Concentration 1.* Click on it.
- Select the left boundary on the graph. Click on the "+" sign at the top right corner, this will add *boundary 1* in the *Boundary Selection* tab. At this point, put $c_{0,c\_A}$: $c_{in}$. This step will add an inlet boundary condition of Equation (6.22a).

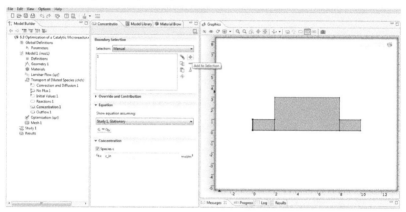

- Again right click on the *Transport of Diluted Species (chds)* tab in *Model Builder.* Select the *Outflow* option. Another window will open namely *Outflow 1.* Click on it.
- Select the right boundary on the graph. Click on the "+" sign at the top right corner, this will add *boundary 12* in the *Boundary Selection* tab. This will add the equation $-n.\ D_i\ \nabla_{ci} = 0$, which is same as the outlet boundary condition as given in Equation (6.22b).

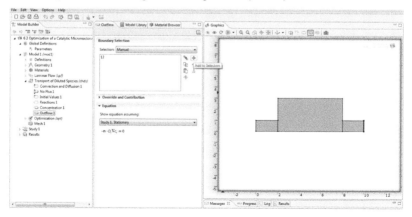

- Again right click on the *Transport of Diluted Species (chds)* tab in *Model Builder.* Select the *Reactions* option. Another window will open namely *Reactions 1.* Click on it.
- Select *Reactions1* option available in the *Transport of Diluted Species (chds)* tab and expand the Reactions tab. Change the $R_c$: – *phi.* Select the *Domain 2* from the graph. Click on the "+" sign at the top right corner, this will add a *Domain 2* in the *Domain Selection* tab.

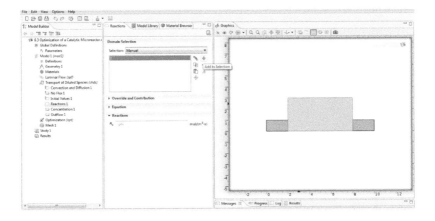

This completes the setup of the physics. Now, set up the optimization problem.

*Step 8*

- Now, solve the optimization problem. Define the control variable epsilon, select its shape, and constrain its values to the interval [0, 1].
- Right click on *Optimization* in *Model Builder* and choose *Control Variable Field.* In the *Control Variable* section choose *Control variable name: epsilon* and *Initial value: 1.*
- In the *Discretization* section, choose *Shape function type: Lagrange* and *Element order: Linear.*
- Select the *Domain 2* from the graph. Click on the "+" sign at the top right corner, this will add a *Domain 2* in the *Domain Selection* tab.

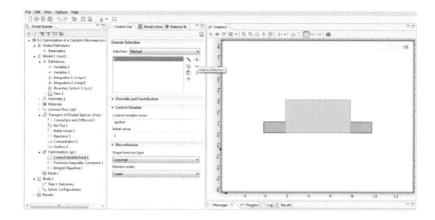

- Right click on *Optimization* in *Model Builder* and choose *Pointwise Inequality Constraint*. In the *Pointwise Inequality Constraint* window choose *Constraint expression: epsilon* and *Bounds: 0 to 1*.
- Select the *Domain 2* from the graph. Click on the "+" sign at the top right corner, this will add a *Domain 2* in the *Domain Selection* tab.

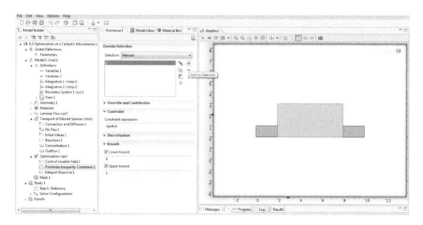

- Next, define the objective function. Right click on *Optimization* in *Model Builder* and choose *Integral Objective*. In the *Integral Objective* window choose *Objective expression: -phi/vol*.
- Select the *Domain 2* from the graph. Click on the "+" sign at the top right corner, this will add a *Domain 2* in the *Domain Selection* tab.

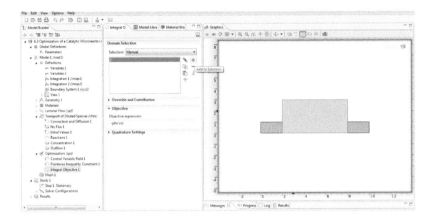

**Step 9**

- Click on the *Mesh* option in *Model Builder.* Select *Fine Mesh* Type.
  Click on *Build All* option at the top of ribbon. A dialogue box will
  appear in the *Message* tab as: *"Complete mesh consists of 12869
  elements."*

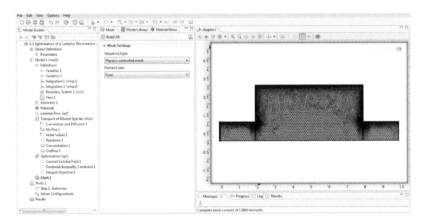

- Now, go to *Study 1* option in the model pellet tab. Click on the
  *Compute* (=) button. A graph will appear giving the profile of velocity
  field in the reactor (see Figure 6.9). This plot is for un-optimized
  solution.

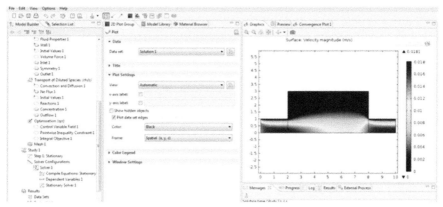

**FIGURE 6.9** Velocity field in the open channel.

*Step 10*

- Now, solve the optimization problem.
- Go to *Stationary* tab under *Study 1* in *Model Builder.* Expand the *Study Extensions* and select *Optimization* check box.

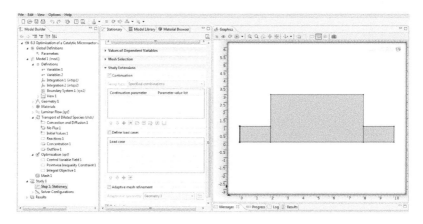

- With the *Model Builder*, right click on *Study 1* option and choose *Show Default Solver.* In the *Stationary 1 node* of *Optimization Solver 1*, change the *Relative tolerance* to 1e-6. Click on the *Compute (=)* button. The velocity in the reactor after optimization should resemble that in Figure 6.10.

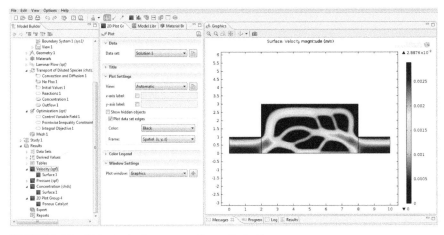

**FIGURE 6.10**    Velocity in the reactor after optimization.

- The concentration distribution in the reactor after optimization should resemble that in Figure 6.11.

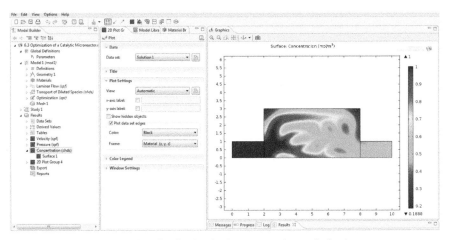

**FIGURE 6.11**    Concentration distribution in the reactor after optimization.

*Step 11*

- To calculate the conversion rate, select *Derived Values* option available in the *Results* tab.
- Now, right click on the *Derived Values* tab and select the *Global Evaluation* option. Another window will open namely *Global Evaluation 1.* Click on it. In the Expression section set *Expression:X.*

- Click on "= Evaluate" button at the top. The conversion of the reactant is around 50% (0.49006).

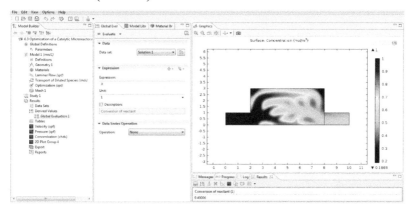

## 6.4 SIMULATION FOR DETERMINING ARRHENIUS PARAMETERS USING PARAMETER ESTIMATION

### 6.4.1 PROBLEM STATEMENT

This model shows how to use the Parameter Estimation feature in the Reaction Engineering physics interface to find the Arrhenius parameters of a first-order reaction.

**Note:** This model requires the optimization module.

### 6.4.2 MODEL DESCRIPTION

Benzene diazonium chloride in the gas phase decomposes to benzene chloride and nitrogen according to

$$
\text{N}\equiv\text{N}-\text{Cl} \quad \xrightarrow{k} \quad \text{Cl} \quad + \text{N}_2 \tag{6.33}
$$

The reaction is first order with the rate

$$
r = kc_{PhN2Cl} \tag{6.34}
$$

where the temperature dependent rate constant given by

$$k = A\exp\left(-\frac{E}{R_g T}\right). \tag{6.35}$$

In Equation (6.35), $A$ and $E$ are the frequency factor (1/s) and the activation energy (J/mol) respectively. The experiments are carried out at different temperatures in a perfectly mixed isothermal batch system to determine the Arrhenius parameters, $A$ and $E$.

The frequency factor $A$ is defined by Equation (6.37) so that the model experiences similar sensitivity with respect to changes in parameter values

$$k = \exp\left(A_{ex}\right)\exp\left(-\frac{E}{R_g T}\right). \tag{6.36}$$

The frequency factor $A$ is then evaluated as

$$A = \exp\left(A_{ex}\right). \tag{6.37}$$

The data indicates that the rate constant is of the order $\sim 1 \times 10^{-3}$ (1/s) at $T = 323$ K. Taking this into account and using an initial guess for the activation energy of 150 kJ/mol, an initial guess is set for $A_{ex} = 49$.

### 6.4.3   SIMULATION APPROACH

*Step 1*

- Open COMSOL Multiphysics.
- Select *0D Space Dimension* from the list of options. Hit the *next* arrow at the upper right corner.

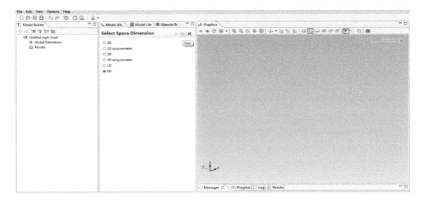

- Expand the *Chemical Species Transport* folder from the list of options in *Model Wizard* and select *Reaction Engineering*. Hit the *next* arrow again.

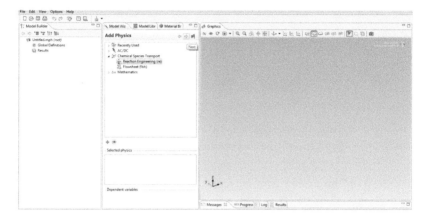

- Then select *Time Dependent* from the list of *Study Type* options, and click on the *Finish flag*.

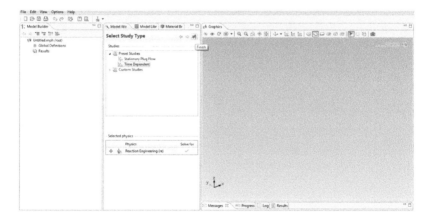

*Step 2*

- With the *Model Builder*, right click on the *Global Definitions* tab and select *Parameters*. Another window will open namely *Parameters*. Click on it and define $T_{iso}$: *313 K*.

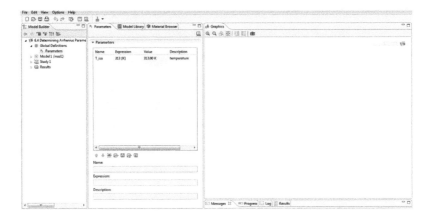

- With the *Model Builder*, right click on the *Reaction Engineering* tab and select the *Reaction* option. Another window will open namely *Reaction 1*. Click on it.
- Select *Reaction 1* option available in the *Reaction Engineering* tab and in the formula edit window type, $PhN_2Cl \Rightarrow PhCl + N_2$. (Note: => Equal to followed by greater sign).
- In the Rate Constants section, check the box "Use Arrhenius expressions" and set $A^f$: $\exp(A_{ex})$, $n^f$: 0 and $E^f$: $E$.

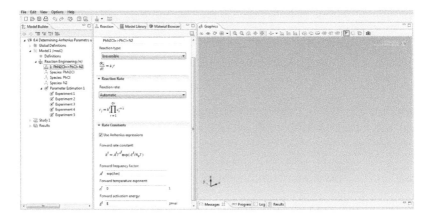

- Select *Species: PhN₂Cl* available in the *Reaction Engineering* tab and in the *General Expression* section type, $c_0$: *1000*.

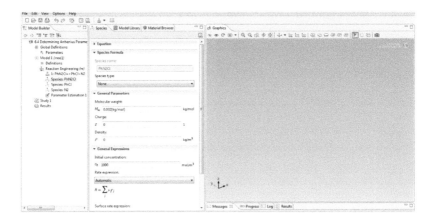

*Step 3*

- With the *Model Builder*, again right click on the *Reaction Engineering* tab in *Model Builder* and select the *Parameter Estimation* option. Another window will open namely *Parameter Estimation 1.* Click on it and enter the variables $A_{ex}$: *49* and *E: 150e3*.

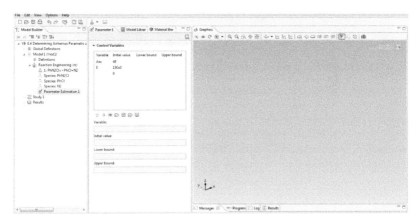

- Right click on the *Parameter Estimation 1* tab in *Model Builder* and select *Experiment* option. Another window will open namely *Experiment 1.*
- Generate the file activation_energy_experiment313K.csv as *comma separated value* files (*csv*-files) with the data given in Table 6.2. Browse the file and click on *Import* button.

**TABLE 6.2** Experimental Data at 313 K

| Time | conc PhN$_2$Cl (313 K) |
|------|------------------------|
| 500  | 799 |
| 1000 | 632 |
| 1500 | 469 |
| 2000 | 376 |
| 2500 | 327 |
| 3000 | 229 |
| 3500 | 186 |
| 4000 | 157 |
| 4500 | 119 |
| 5000 | 83 |

- In the table, set the Model variables *time: t* and conc PhN$_2$Cl (313 K): $c_{PhN2Cl}$.
- In the Experimental Parameters section, click on "+" button at the bottom and enter *Parameter names: T$_{iso}$* and *Parameter expressions: 313*.

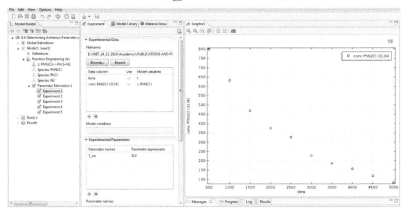

*Step 4*

- Again right click on the *Parameter Estimation 1* tab in *Model Builder* and select *Experiment* option. Another window will open namely *Experiment 2*.
- Generate the file activation_energy_experiment319K.csv as *comma separated value* files (*csv*-files) with the data given in Table 6.3. Browse the file and click on *Import* button.

**TABLE 6.3**   Experimental Data at 319 K

| Time | conc PhN$_2$Cl (319 K) |
|------|------------------------|
| 300  | 736 |
| 600  | 513 |
| 900  | 344 |
| 1200 | 285 |
| 1500 | 206 |
| 1800 | 141 |
| 2100 | 91  |
| 2400 | 70  |
| 2700 | 42  |
| 3000 | 34  |

- In the table, set the Model variables *time: t* and conc PhN$_2$Cl (319 K): $c_{PhN2Cl}$.
- In the Experimental Parameters section, click on "+" button at the bottom and enter *Parameter names: T$_{iso}$* and *Parameter expressions: 319.*

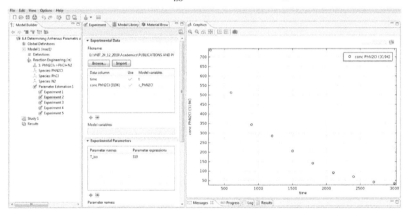

*Step 5*

- Again right click on the *Parameter Estimation 1* tab in *Model Builder* and select *Experiment* option. Another window will open namely *Experiment 3.*
- Generate the file activation_energy_experiment323K.csv as *comma separated value* files (*csv*-files) with the data given in Table 6.4. Browse the file and click on *Import* button.

**TABLE 6.4**   Experimental Data at 323 K

| Time | conc PhN$_2$Cl (323 K) |
|------|------------------------|
| 150  | 772 |
| 300  | 546 |
| 450  | 420 |
| 600  | 342 |
| 750  | 256 |
| 900  | 170 |
| 1050 | 128 |
| 1200 | 96 |
| 1350 | 81 |
| 1500 | 65 |

- In the table, set the Model variables *time: t* and conc PhN$_2$Cl (323 K): $c_{PhN2Cl}$.
- In the Experimental Parameters section, click on "+" button at the bottom and enter *Parameter names: T$_{iso}$* and *Parameter expressions: 323.*

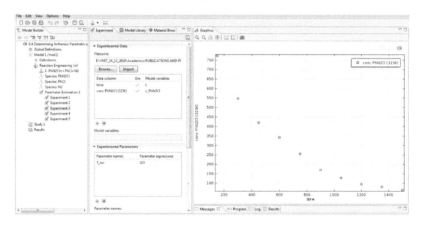

*Step 6*

- Again right click on the *Parameter Estimation 1* tab in *Model Builder* and select *Experiment* option. Another window will open namely *Experiment 4.*
- Generate the file activation_energy_experiment328K.csv as *comma separated value* files (*csv*-files) with the data given in Table 6.5. Browse the file and click on *Import* button.

**TABLE 6.5**  Experimental Data at 328 K

| Time | conc PhN$_2$Cl (328 K) |
|------|------------------------|
| 50   | 841 |
| 100  | 667 |
| 150  | 554 |
| 200  | 458 |
| 250  | 394 |
| 300  | 326 |
| 350  | 282 |
| 400  | 225 |
| 450  | 195 |
| 500  | 166 |

- In the table, set the Model variables *time: t* and conc PhN$_2$Cl (*328* K): $c_{PhN2Cl}$.
- In the Experimental Parameters section, click on "+" button at the bottom and enter *Parameter names: T$_{iso}$* and *Parameter expressions: 328*.

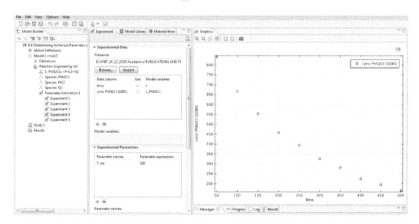

*Step 7*

- Again right click on the *Parameter Estimation 1* tab in *Model Builder* and select *Experiment* option. Another window will open namely *Experiment 5*.

- Generate the file activation_energy_experiment333K.csv as *comma separated value* files (*csv*-files) with the data given in Table 6.6. Browse the file and click on *Import* button.

**TABLE 6.6**    Experimental Data at 333 K

| Time | conc PhN$_2$Cl (333 K) |
|------|------------------------|
| 30   | 804                    |
| 60   | 660                    |
| 90   | 550                    |
| 120  | 444                    |
| 150  | 373                    |
| 180  | 283                    |
| 210  | 239                    |
| 240  | 186                    |
| 270  | 152                    |
| 300  | 120                    |

- In the table, set the Model variables *time: t* and conc PhN$_2$Cl (333 K): $c_{PhN2Cl}$.
- In the Experimental Parameters section, click on "+" button at the bottom and enter *Parameter names: T$_{iso}$* and *Parameter expressions: 333*.

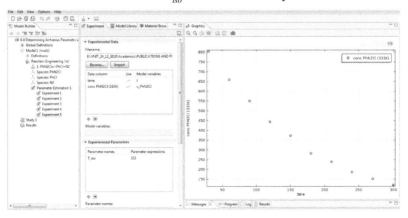

*Step 8*

- In the *Reaction Engineering* tab in *Model Builder*, expand the *General section* and set *T: T$_{iso}$*.

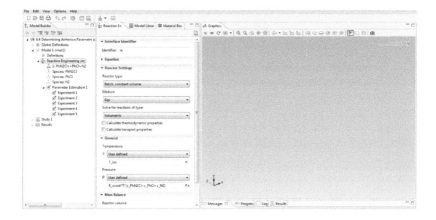

## Step 9

- Now, go to *Study 1, Step1: Time Dependent* option in the model pellet tab and set the *Times: range (0,50,5000)*.

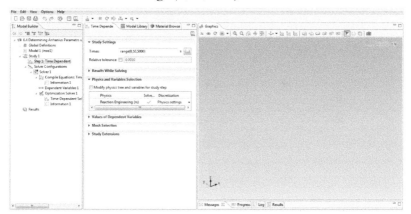

## Step 10

- Right click on the *Study 1* tab in *Model Builder* and select *Show Default Solver* option. Go to *Solver 1* under *Solver Configurations*. Right click on it and select *Optimization Solver 1*.
- Then right click on *Optimization Solver 1* node, and select *Time-Dependent Solver 1*. Remove *Time-Dependent Solver 1* if it is present under Solver 1.
- In the *Optimization Solver 1* window, from the *Method* list, choose *Levenberg–Marquardt*.

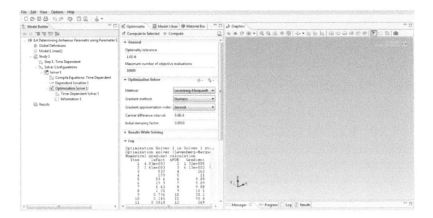

## Step 11

- Go to *Time-Dependent Solver 1* window under *Optimization Solver 1*. Expand the *Absolute tolerance* section and choose *Global method: Unscaled* and *Tolerance: 1e-5*. Expand the *Output* section and choose *Times to store: Specified values*. Click on the *Compute* (=) button.

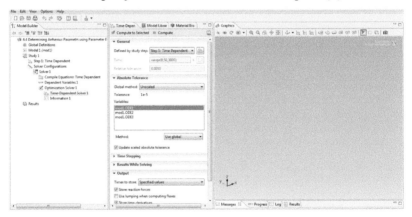

## Step 12

- In the *Model Builder*, expand *Experiment 1 Group node* available under *Results* and select *Global 1* option. In the *Global 1* window, expand the Data section. Set *Parameter selection ($T_{iso}$): From list* and *Parameter values ($T_{iso}$): 313*.
- In the *y*-axis data section, select the *Expression: mod1.re.c_PhN$_2$Cl* and in the *x*-axis data section, select the select the *Axis source data: Time*. Click on the *Plot* at the top.

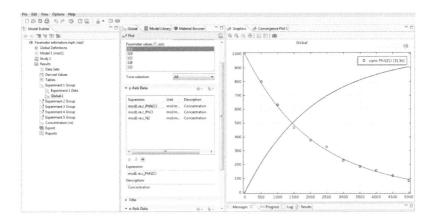

*Step 13*

- In the *Model Builder*, expand *Experiment 1 Group node* available under *Results* and select *Global 1* option. In the *Global 1* window, expand the Data section. Set *Parameter selection (T$_{iso}$): From list* and *Parameter values (T$_{iso}$): 319.*
- In the *y*-axis data section, select the *Expression: mod1.re.c_PhN$_2$Cl* and in the *x*-axis data section, select the select the *Axis source data: Time*. Click on the *Plot* at the top.

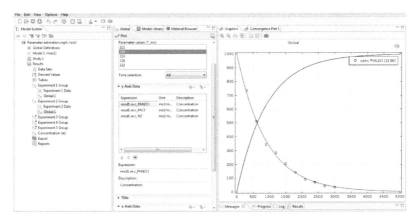

*Step 14*

- In the *Model Builder*, expand *Experiment 1 Group node* available under *Results* and select *Global 1* option. In the *Global 1* window, expand the Data section. Set *Parameter selection ($T_{iso}$): From list* and *Parameter values ($T_{iso}$): 323.*

- In the *y*-axis data section, select the *Expression: mod1.re.c$_{PhN2Cl}$* and in the *x*-axis data section, select the select the *Axis source data: Time.* Click on the *Plot* at the top.

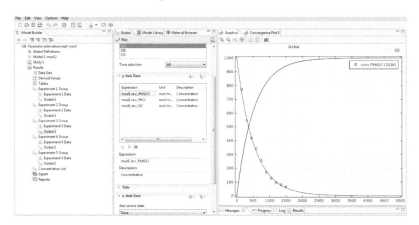

*Step 15*

- In the *Model Builder*, expand *Experiment 1 Group node* available under *Results* and select *Global 1* option. In the *Global 1* window, expand the Data section. Set *Parameter selection ($T_{iso}$): From list* and *Parameter values ($T_{iso}$): 328.*
- In the y-axis data section, select the *Expression: mod1.re.c$_{PhN2Cl}$* and in the *x*-axis data section, select the select the *Axis source data: Time.* Click on the *Plot* at the top.

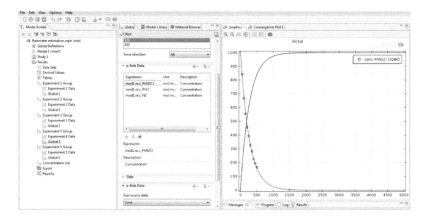

*Step 16*

- In the *Model Builder*, expand *Experiment 1 Group node* available under *Results* and select *Global 1* option. In the *Global 1* window, expand the Data section. Set *Parameter selection ($T_{iso}$): From list* and *Parameter values ($T_{iso}$): 333*.
- In the *y*-axis data section, select the *Expression: mod1.re.c$_{PhN2Cl}$* and in the *x*-axis data section, select the select the *Axis source data: Time*. Click on the *Plot* at the top.

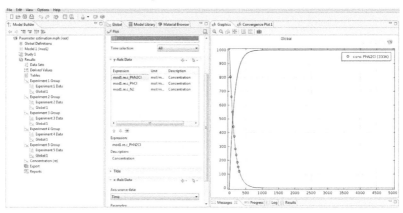

*Step 17*

- In the *Results tab*, right click on *Derived Values* and click *Evaluate All*. *E* is found to be 1.16e5 (J/mol) and $A_{ex}$ is evaluated to 36.93.
- From Equation (6.37), we have

$$A = \exp(A_{ex})$$

A = exp (36.93) = $1.093 \times 10^{16}$ (1/s).

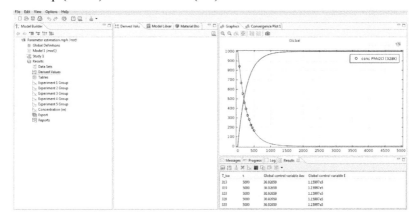

## 6.5 PROBLEMS

1. A chemical company produces three different products *A, B,* and *C.* The products are produced at company's two different facilities having different production capacities. In a normal 8-h working day, plant 1 produces 100, 200, and 200 kg of products *A, B,* and *C.* Plant 2 produces 120, 120, and 400 kg of *A, B,* and *C,* respectively. The daily cost of production for plant 1 and 2 is Rs. 5000 and 7000, respectively. Find the minimum number of days of operation at two different plants to minimize the total cost while meeting the demand.

2. The manufacturer wish to blend three metals *A, B,* and *C* to form 10 tons of an alloy. The alloy must satisfy certain specifications. The alloy must contain at least 25% lead, not more than 50% tin and at least 20% zinc. The composition and costs of the three metals are shown in the table

| Metal | A | B | C |
|---|---|---|---|
| Lead | 0.10 | 0.10 | 0.40 |
| Tin | 0.10 | 0.30 | 0.60 |
| Zinc | 0.80 | 0.60 | 0.00 |
| Rs/ton | 1400 | 2000 | 3000 |

What blend of these metal produces alloys which satisfy the specification at minimum costs.

3. The soft drink plant has two bottling machines "I" and "II." "I" is designed for 10-ounce bottles and "II" for 18-ounce bottles. Each machine can be used on any type of bottle with some loss in production. The following table gives the data on each type of machine:

| Machine | 10-ounce bottles | 18-ounce bottles |
|---|---|---|
| I | $110 \ min^{-1}$ | $50 \ min^{-1}$ |
| II | $70 \ min^{-1}$ | $85 \ min^{-1}$ |

The profit in a 10-ounce bottle is 25 paise and on 18-ounce bottle is 35 paise. Weekly production of drink cannot exceed 350,000 ounces and the market can absorb 30,000, 10-ounce bottles and 10,000, 18-ounce bottles per week. The planner wishes to maximize his profit subject to all the production and marketing restrictions. The machines can run 8 h per day, 5 days per week.

4. Maximum $Z = 2x + 3y$

   The constraints are

   $x + y \leq 30,\ y \geq 3$

   $0 \leq y \leq 12,\ x - y \geq 0$

   $0 \leq x \leq 20.$

5. Three different colors of wool are used by the company to produce two types of clothes. Two yard length of type I cloth requires 8 oz of blue wool, 10 oz of pink wool, and 6 oz of yellow wool. Two yard length of type II cloth requires 10 oz of blue wool, 4 oz of pink wool, and 16 oz of yellow wool. The company has 2000 oz of blue wool, 2000 oz of pink wool, and 2400 oz of yellow wool. The company can make a profit of Rs. 10 on one yard of type I cloth and Rs. 6 on one yard of type II cloth. Find the best combination of the quantities of type I and type II cloth which gives him maximum profit.

6. Minimize the function $f(x) = 100(x_2 - x_1^2)^2 + (1 - x_1)^2$

   subject to: $x_1^2 + x_2^2 \leq 2.$

7. Consider the function:

$$C = \$500 + \$0.9x + \frac{\$0.03}{x}(150,000).$$

   $C$ is the total annual cost of operating a pump and motor and $x$ is size (horsepower) of the motor.

   Find the motor size that minimizes the total annual cost.

8. The filtration is carried out as per the equation given by Cook (1984):

$$\text{Flow rate} \propto \frac{(\text{Pressue drop})(\text{Filter area})}{(\text{Fluid viscosity})(\text{Cake thickness})}.$$

   Cook expressed filtration time as

$$t_f = \beta \frac{\Delta P_c A^2}{\mu M^2 c} x_c \exp(-ax_c + b)$$

   where

   $t_f$ is the time to build filter cake, min; $\beta$ is $3.2 \times 10^{-8}$ $(\text{lb}_m/\text{ft})^2$; $DP_c$ is the pressure drop across cake, psig (20); $A$ is the filtration area (250 ft²); $x_c$ is the mass fraction solids in dry cake; $a$ is the constant relating cake resistance to solids fraction (3.643); $b$ is the constant relating cake resistance to solids fraction (2.680); is the filtrate viscosity, centipoises (20); $M$ is the mass flow of the filtrate, (75 $\text{lb}_m/\text{min}$); and

$c$ is the solid concentration in feed to filter, $lb_m/lb_m$ filtrate (0.01). Obtain the maximum time for filtratin as a function of $x_c$.

9. In a decision problem, it is desired to minimize the expected risk defined as follows:

$$\varepsilon(\text{risk}) = (1-P)c_1\left[1-F(b)\right] + Pc_2\theta\left(\frac{b}{2} + \frac{2\pi}{4}\right)F\left(\frac{b}{2} - \frac{\sqrt{2\pi}}{4}\right)$$

where

$$F(b) = \int_{-\infty}^{b} e^{-u^2/2\theta^2}\,du \quad \text{(normal probability function)}$$

$c_1 = 1.25 \times 10^5$
$c_2 = 15$
$\theta = 2000$
$P = 0.25.$

Find the minimum expected risk and $b$.

10. Solve the following problem (Edgar et al., 2001):
Maximize: $f = 7x_1 + 12x_2 + 3x_3$
subject to: $2x_1 + 2x_2 + x_3 \le 16$
$\qquad\qquad 4x_1 + 8x_2 + x_3 \le 40$
$\qquad\qquad x_1, x_2, x_3 \ge 0.$

# CHAPTER 7

# Case Studies

## 7.1 CASE STUDY ON MODELING, SIMULATION, AND OPTIMIZATION OF DOWNDRAFT GASIFIER

Biomass is a potentially sustainable, renewable, and relatively environmentally acceptable source of energy. Pyrolysis converts biomass into a char, volatiles, and gases. Tar forms a major part of the primary volatile yield and readily undergoes secondary reactions, which are crucial in gasification and combustion technologies. This case study proposes a more sophisticated 1D steady-state model of a downdraft gasifier.

### 7.1.1 MODEL DEVELOPMENT FOR DOWNDRAFT GASIFIER

*Gas phase species*

$$v_g \frac{dC_{CO}}{dz} = -C_{CO} \frac{dv_g}{dz} + r_{1(c)} - 2r_{3(c)} - r_{shift} + 2r_{1(r)} + r_{2(r)} + \overline{\gamma}_{CO} r_{T,crack} \quad (7.1)$$

$$v_g \frac{dC_{CO_2}}{dz} = -C_{CO_2} \frac{dv_g}{dz} + 2r_{3(c)} + r_{shift} - r_{1(r)} + \overline{\gamma}_{CO_2} r_{T,crack} \quad (7.2)$$

$$v_g \frac{dC_{H_2}}{dz} = -C_{H_2} \frac{dv_g}{dz} - 2r_{2(c)} + r_{shift} + r_{2(r)} - 2r_{3(r)} \quad (7.3)$$

$$v_g \frac{dC_{CH_4}}{dz} = -C_{CH_4} \frac{dv_g}{dz} - r_{1(c)} + r_{3(r)} + \overline{\gamma}_{CH_4} r_{T,crack} \quad (7.4)$$

$$v_g \frac{dC_{H_2O}}{dz} = -C_{H_2O} \frac{dv_g}{dz} + 2r_{1(c)} + 2r_{2(c)} - r_{shift} - r_{2(r)} \quad (7.5)$$

$$v_g \frac{dC_T}{dz} = -C_T \frac{dv_g}{dz} - r_{T,crack} \quad (7.6)$$

$$v_g \frac{dC_{O_2}}{dz} = -C_{O_2} \frac{dv_g}{dz} - 1.5r_{1(c)} - r_{2(c)} - r_{3(c)} \quad (7.7)$$

$$v_g \frac{dC_{N_2}}{dz} = -C_{N_2} \frac{dv_g}{dz} \qquad (7.8)$$

Energy balance:

$$\left(v_g \sum_i C_i c_{pi}\right) \frac{dT_g}{dz} = -\sum_j r_j \Delta H_j - v_g \frac{dP}{dz} - P \frac{dv_g}{dz} - \sum_i R_i' c_{pi} T_g - \frac{4h_{gw}}{D_r}(T_g - T_w) \;(7.9)$$

$$-\frac{dP}{dz} = \frac{150\mu(1-\varepsilon)^2}{d_p^2 \varepsilon^3} v_g + \frac{1.75 C_g (1-\varepsilon)}{d_p \varepsilon^3} v_g^2 \qquad (7.10)$$

$$\frac{dv_g}{dz} = \frac{1}{C_g R_g + \sum_i C_i c_{pi}} \left[ -\frac{\sum_j r_j \Delta H_j}{T_g} - \frac{dP}{dz}\left(\frac{v_g}{T_g} + \frac{v_g \sum_i C_i c_{pi}}{P} + \frac{\sum_i C_i c_{pi} \sum_i R_i'}{C_g}\right) \right.$$
$$\left. -\sum_i R_i' c_{pi} - \frac{4h_{gw}(T_g - T_w)}{D_r T_g} \right] \qquad (7.11)$$

Equations (7.1)–(7.11) were solved using COMSOL Multiphysics software. The correlations and initial values used to solve the model are available in earlier research work (Chaurasia, 2016).

### 7.1.2  SIMULATION APPROACH

*Step 1*

- Select *0D Space Dimension* from the list of options by opening the COMSOL Multiphysics. Hit the *next* arrow at the upper right corner.

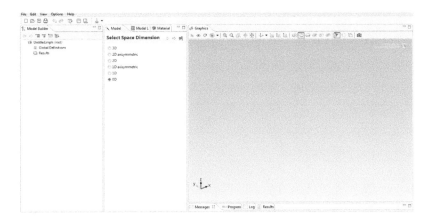

- Select and expand the *Mathematics* folder from the list of options in *Model Wizard*. Further choose *Global ODEs and DAEs* and hit *next tab*.

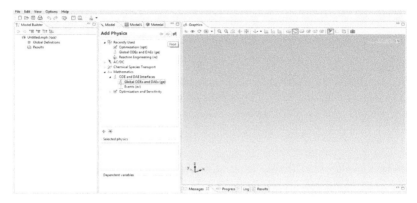

- Then, select *Time Dependent* from the list of *Study Type* options, and click on the *Finish flag* at the upper right corner of the application.

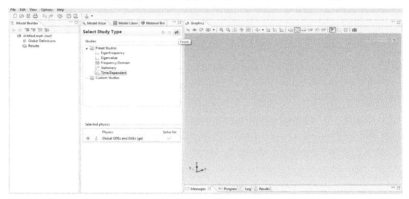

*Step 2*

- Now, in the *Model Builder*, select *Global Equations* option. This will open a tab to enter the coefficients of the characteristics of a model equation.
- In the *Global Equations* panel, you will see an equation of the form

$$f\left(u, u_t, u_{tt}, t\right) = 0, u\left(t_0\right) = u_0, u_t\left(t_0\right) = u_{t0}. \tag{7.12}$$

To convert Equations (7.1)–(7.11) to the desired form of Equation (7.12), enter the values as given in Table 7.1.

**TABLE 7.1** Equations and Initial Conditions

| Name | $f(u,u_t,u_{tt},t)$ | Initial value ($u_0$) | Initial value ($u_{t0}$) | Description |
|---|---|---|---|---|
| $pCO$ | $pCO_t - (1/v)*$ $(RCO - pCO*(vt))$ | 0.6*0.1697 | 0 | Molar concentration of CO on moisture free basis (moles/m$^3$) |
| $pCO_2$ | $pCO_2t - (1/v)*$ $(RCO_2 - pCO_2*(vt))$ | 0.6*0.217 | 0 | Molar concentration of $CO_2$ on moisture free basis (moles/m$^3$) |
| $pH_2$ | $pH_2t - (1/v)*(RH_2 - pH_2*(vt))$ | 0.6*0.0213 | 0 | Molar concentration of $H_2$ on moisture free basis (moles/m$^3$) |
| $pCH_4$ | $pCH_4t - (1/v)*(RCH_4$ $- pCH_4*(vt))$ | 0.6*0.072 | 0 | Molar concentration of $CH_4$ on moisture free basis (moles/m$^3$) |
| $pH_2O$ | $pH_2Ot - (1/v)*$ $(RH_2O - pH_2O*(vt))$ | 0.6*0.282 | 0 | Molar concentration of $CH_4$ on moisture free basis (moles/m$^3$) |
| $pN_2$ | $pN_2t - (1/v)*(RN_2 - pN_2*(vt))$ | 0.79 | 0 | Molar concentration of $N_2$ on moisture free basis (moles/m$^3$) |
| $pO_2$ | $pO_2t - (1/v)*(RO_2 - pO_2*(vt))$ | 0.21 | 0 | Molar concentration of $O_2$ on moisture free basis (moles/m$^3$) |
| $pT$ | $pTt - (1/v)*(RTar - pT*(vt))$ | 0.6*0.24 | 0 | Molar concentration of tar on moisture free basis (moles/m$^3$) |
| $v$ | $vt - (1/(A + pg*Rg))*(((A*D)/$ $pg) - (B/T) - Pt*(v/T +$ $(v*A)/P) - C - EE/T)$ | 0.699 | 0 | Superficial gas velocity (m/s) |
| $T$ | $Tt - (1/(v*A))*(-B - v*$ $Pt - P*vt - C*T - EE)$ | 1228 | 0 | Temperatue of gas (K) |
| $P$ | $-Pt - (150*mu*(1 - e)^2*v)/$ $(dp*2*e^3) + (1.75 + pg*$ $(1 - e)*v^2)/(dp*e^3)$ | 101325 | 0 | Total pressure of species (Pa) |

The initial values of different variables at $t = 0$ and COMSOL notations of Equations (7.1) – (7.11) are indicated in the above table.

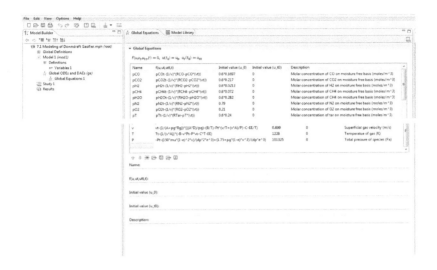

## Step 3

- In the *Model Builder*, select *Variables* by right clicking on the *Definitions* tab. Another window will open namely *Variables 1*. Click on it and define the variables as given in Table 7.2.

**TABLE 7.2** Variables and Expressions in the Model Equations

| Name | Expression | Description |
|---|---|---|
| $pg$ | $pCO+pCO_2+pH_2+pCH_4$ $+pH_2O+pN_2+pO2+pT$ | Molar concentration of all gaseous species on moisture free basis (moles/m³) |
| $pgdb$ | $pCO+pCO_2+pH_2+pCH_4$ $+pN_2+pO_2+pT$ | Molar concentration of all gaseous species on dry basis (moles/m³) |
| $pgm$ | $pg*Mw$ | Mass concentration of all gaseous species (kg/m³) |
| $xCO$ | $(pCO/pg)*100$ | Molar fraction of CO on moisture basis (vol%) |
| $xCO_2$ | $(pCO_2/pg)*100$ | Molar fraction of $CO_2$ on moisture basis (vol%) |
| $xH_2$ | $(pH_2/pg)*100$ | Molar fraction of $H_2$ on moisture basis (vol%) |
| $xCH_4$ | $(pCH4/pg)*100$ | Molar fraction of $CH_4$ on moisture basis (vol%) |
| $xH_2O$ | $(pH_2O/pg)*100$ | Molar fraction of $H_2O$ on moisture basis (vol%) |
| $xN_2$ | $(pN_2/pg)*100$ | Molar fraction of $N_2$ on moisture basis (vol%) |
| $xO_2$ | $(pO_2/pg)*100$ | Molar fraction of $O_2$ on moisture basis (vol%) |
| $xT$ | $(pT/pg)*100$ | Molar fraction of tar on moisture basis (vol%) |

**TABLE 7.2** *(Continued)*

| Name | Expression | Description |
|---|---|---|
| $xCOdb$ | $(pCO/pgdb)*100$ | Molar fraction of CO on dry basis (vol%) |
| $xCO_2db$ | $(pCO_2/pgdb)*100$ | Molar fraction of $CO_2$ on dry basis (vol%) |
| $xH_2db$ | $(pH_2/pgdb)*100$ | Molar fraction of $H_2$ on dry basis (vol%) |
| $xCH_4db$ | $(pCH4/pgdb)*100$ | Molar fraction of $CH_4$ on dry basis (vol%) |
| $xN_2db$ | $(pN2/pgdb)*100$ | Molar fraction of $N_2$ on dry basis (vol%) |
| $xO_2db$ | $(pO_2/pgdb)*100$ | Molar fraction of $O_2$ on dry basis (vol%) |
| $xTdb$ | $(pT/pgdb)*100$ | Molar fraction of tar on dry basis (vol%) |
| ratio | $xCOdb/xCO_2db$ | Ratio of CO to $CO_2$ (–) |
| $k_1r$ | $A_1r*\exp(-E_1r/(Rg*T))$ | Rate constant of reduction reaction 1 (1 s$^{-1}$) |
| $k_2r$ | $A_2r*\exp(-E_2r/(Rg*T))$ | Rate constant of reduction reaction 2 (1 s$^{-1}$) |
| $k_3r$ | $A_3r*\exp(-E_3r/(Rg*T))$ | Rate constant of reduction reaction 3 (1 s$^{-1}$) |
| $k_1c$ | $A_1c*\exp(-E_1c/(Rg*T))$ | Rate constant of combustion reaction 1 (1 s$^{-1}$) |
| $k_2c$ | $A_2c*\exp(-E_2c/(Rg*T))$ | Rate constant of combustion reaction 2 (1 s$^{-1}$) |
| $k_3c$ | $A_3c*\exp(-E_3c/(Rg*T))$ | Rate constant of combustion reaction 3 (1 s$^{-1}$) |
| kshift | $Ashift*\exp(-Eshift/(Rg*T))$ | Rate constant of shift reaction (moles/m$^3$.s) |
| keq | $0.0265*\exp(7914/T)$ | Equilibrium constant (–) |
| kT | $AT*\exp(-ET/(Rg*T))$ | Rate constant of tar cracking reaction (moles/m$^3$.s) |
| ap | $(6*(1-e))/(dp)$ | Volumetric solid surface area (m$^2$/m$^3$) |
| mu | $(1.98e-5)*((T/300)^{(2.0/3.0)})$ | Viscosity of gas (kg/m.s) |
| Re | $(pgm*dp*v)/mu$ | Reynolds number (–) |
| Sc | $mu/(pgm*Di)$ | Schmidt number (–) |
| Pr | $(mu*cg)/kg$ | Prandtl number (–) |
| km | $((2.06*v)/e)*(1.0/(Re^{0.575}))*(1.0/(Sc^{(2.0/3.0)}))$ | Mass transfer coefficient (m/s) |
| hsg | $((corr*2.06*cg*pg*v)/e)*(1.0/(Re^0.575))*(1.0/(Sc^{(2.0/3.0)}))$ | Heat transfer coefficient (W/m$^2$K) |
| $r_1r$ | $(pCO_2)/((1.0/(km*ap))+(1.0/k_1r))$ | Rate constant of reduction reaction 1 (moles/m$^3$.s) |
| $r_2r$ | $(pH_2O)/((1.0/(km*ap))+(1.0/k_2r))$ | Rate constant of reduction reaction 2 (moles/m$^3$.s) |
| $r_3r$ | $(pH_2)/((1.0/(km*ap))+(1.0/k_3r))$ | Rate constant of reduction reaction 3 (moles/m$^3$.s) |
| $r_1c$ | $k_1c*(pCH_4^{0.7})*(pO_2^{0.8})$ | Rate constant of combustion reaction 1 (moles/m$^3$.s) |
| $r_2c$ | $k2c*pH2*pO2$ | Rate constant of combustion reaction 2 (moles/m$^3$.s) |
| $r_3c$ | $k_3c*pCO*pO_2*(pH_2O^{0.5})$ | Rate constant of combustion reaction 3 (moles/m$^3$.s) |

**TABLE 7.2** *(Continued)*

| Name | Expression | Description |
|---|---|---|
| $r$shift | $k$shift$*((pCO*pH_2O)$ $-(pCO* pH_2/keq))$ | Rate constant of shift reaction (moles/m³.s) |
| $rT$ | $kT*pT$ | Rate constant of tar cracking reaction (moles/m³.s) |
| RCO | $2.0*r_1r+r_2r+vCO*rT+r_1c-$ $2.0*r_3c-r$shift | Net rate of creation of CO by chemical reactions (moles/m³.s) |
| $RCO_2$ | $-r_1r+vCO_2*rT+2.0*r_3c+r$shift | Net rate of creation of $CO_2$ by chemical reactions (moles/m³.s) |
| $RH_2$ | $r_2r+vH_2*rT-2.0*r_3r-$ $2.0*r_2c+r$shift | Net rate of creation of $H_2$ by chemical reactions (moles/m³.s) |
| $RCH_4$ | $r_3r+vCH_4*rT-r_1c$ | Net rate of creation of $CH_4$ by chemical reactions (moles/m³.s) |
| $RH_2O$ | $-r_2r+2.0*r_1c+2.0*r_2c-r$shift | Net rate of creation of $H_2O$ by chemical reactions (moles/m³.s) |
| $RN_2$ | $0$ | Net rate of creation of $N_2$ by chemical reactions (moles/m³.s) |
| $RO_2$ | $-1.5*r_1c-r_2c-r_3c$ | Net rate of creation of $O_2$ by chemical reactions (moles/m³.s) |
| *RTar* | $-rT+vST*rT$ | Net rate of creation of tar by chemical reactions (moles/m³.s) |
| *A* | $pCO*cg+pCO_2*cg+pH_2*cg+p$ $CH_4*cg+pH_2O*cg+pN_2*cg+p$ $O_2*cg+pT*cg$ | Used to calculate superficial velocity (J/m³.K) |
| *B* | $r_1r*dH_1r+r_2r*dH_2r+r_3r*dH_3r+$ $rT*dHT+r$shift$*dH$shift$+r1c*$ $dH_1c+r_2c*dH_2c+r_3c*dH_3c$ | Used to calculate superficial velocity (J/m³.K) |
| *C* | $RCO*cg+RCO_2*cg+RH_2*cg+$ $RCH_4*cg+RH_2O*cg+RN_2*cg$ $+RO_2*cg+rT*cg$ | Used to calculate superficial velocity (J/m³.K) |
| *D* | $RCO+RCO_2+RH_2+RCH_4+RH$ $_2O+RN_2+RO_2+rT$ | Total moles of gas used to calculate superficial velocity (J/m³.K) |
| *EE* | $4.0*hsg*(T-Tw)/Dia$ | Used to calculate temperature (J/s.m³) |
| $A_1r$ | $2.00E+12$ | 1 s⁻¹ |
| $A_2r$ | $2.40E+12$ | 1 s⁻¹ |
| $A_3r$ | $1.59E+04$ | 1 s⁻¹ |
| $A_1c$ | $1.60E+08$ | 1 s⁻¹ |
| $A_2c$ | $3.98E+05$ | 1 s⁻¹ |
| $A_3c$ | $3.70E+09$ | 1 s⁻¹ |
| *Ashift* | $2.78E+04$ | 1 s⁻¹ |

**TABLE 7.2**   *(Continued)*

| Name | Expression | Description |
|------|-----------|-------------|
| $AT$ | 6.31$E$+10 | 1 s$^{-1}$ |
| $E_1r$ | 216953.83 | J/mol |
| $E_2r$ | 216953.83 | J/mol |
| $E_3r$ | 216953.83 | J/mol |
| $E_1c$ | 200841.3 | J/mol |
| $E_2c$ | 83140 | J/mol |
| $E_3c$ | 125524.77 | J/mol |
| $Eshift$ | 12579.1 | J/mol |
| $ET$ | 185000.77 | J/mol |
| $Rg$ | 8.314 | J/mol.K |
| $cg$ | 1100 | J/mol.K |
| $dH_1r$ | −172,600 | J/mol |
| $dH_2r$ | −131,400 | J/mol |
| $dH_3r$ | 75,000 | J/mol |
| $dHT$ | 42,000 | J/mol |
| $dHshift$ | 41,200 | J/mol |
| $dH_1c$ | 17,473 | J/mol |
| $dH_2c$ | 142,919 | J/mol |
| $dH_3c$ | 10,107 | J/mol |
| $e$ | 0.394 | − |
| $dp$ | 0.001 | m |
| $Mw$ | 32.75 | kg/kgmol or g/mol |
| $Di$ | 2.00$E$−05 | m$^2$/s |
| $vCO$ | 0.497 | − |
| $vCO_2$ | 0.322 | − |
| $vCH_4$ | 0.102 | − |
| $vH_2$ | 0.026 | |
| $vST$ | 0.06 | |
| $Dia$ | 0.2 | m |
| $Tw$ | 300 | K |
| $corr$ | 0.02 | |
| $kg$ | 2.58$E$−02 | W/m K |

*Step 4*

- Now, expand the *Study 1* tab in *Model Builder*. Select the *Step 1: Time Dependent*. Another window will open namely *Time Dependent*. Expand the *Study Settings* tab and set the range *Times: range (0,0.02,3)*.

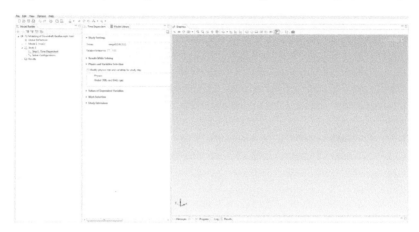

- Click on the *Compute (=)* button. Save the simulation.

*Step 5*

- With the *Model Builder*, right click on the *Global 1* option under 1D Plot Group 1 and select *Add Plot Data to Export*. Another window will open namely *Plot*. Click on it and give the file name with .csv extension. Then, press the *Export* button. This will generate the data as given in Table 7.3 for pyrolysis fraction of 0.6.

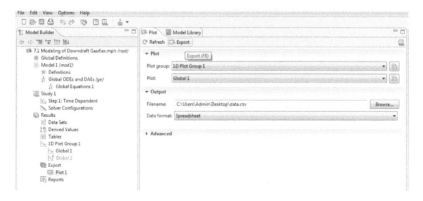

**TABLE 7.3**   Simulation Results With Pyrolysis Fraction of 0.6

| z (m) | xCH₄db | xCO₂db | xCOdb | xH₂db | xN₂db | xO₂db | xTdb | T (K) |
|-------|--------|--------|-------|-------|-------|-------|------|-------|
| *0* | 3.015798 | 10.64951 | 6.976649 | 2.294378 | 54.08009 | 13.66576 | 9.317817 | 1211.422 |
| 0.02 | 3.013699 | 16.98257 | 18.74898 | 10.25669 | 44.56598 | 3.629251 | 2.802815 | 1002.835 |
| 0.04 | 3.027708 | 16.71472 | 21.53756 | 10.15681 | 43.9227 | 2.40864 | 2.231867 | 976.898 |
| 0.06 | 3.030826 | 16.4204 | 23.15419 | 10.09367 | 43.54659 | 1.809452 | 1.944869 | 960.8805 |
| 0.08 | 3.032715 | 16.14221 | 24.52783 | 10.03869 | 43.22696 | 1.323597 | 1.707993 | 946.5014 |
| 0.1 | 3.033493 | 15.87936 | 25.67644 | 9.991273 | 42.95967 | 0.942446 | 1.517311 | 933.7342 |
| 0.12 | 3.032928 | 15.64026 | 26.51747 | 9.956441 | 42.76238 | 0.703444 | 1.387075 | 923.3803 |
| 0.14 | 3.03209 | 15.42117 | 27.23311 | 9.926505 | 42.59421 | 0.513554 | 1.279354 | 913.9015 |
| 0.16 | 3.030993 | 15.22131 | 27.82853 | 9.901254 | 42.45398 | 0.370715 | 1.193212 | 905.3082 |
| 0.18 | 3.029499 | 15.04989 | 28.2581 | 9.883699 | 42.35125 | 0.291093 | 1.136471 | 898.0265 |
| 0.2 | 3.027915 | 14.90521 | 28.5742 | 9.870751 | 42.27497 | 0.248242 | 1.098724 | 891.6339 |
| 0.22 | 3.026577 | 14.79181 | 28.8106 | 9.85986 | 42.21865 | 0.220082 | 1.072416 | 885.8355 |
| 0.24 | 3.025231 | 14.6882 | 29.00699 | 9.853234 | 42.16959 | 0.205597 | 1.051159 | 880.2925 |
| 0.26 | 3.024178 | 14.60632 | 29.16191 | 9.847672 | 42.13117 | 0.194255 | 1.034498 | 875.0255 |
| 0.28 | 3.023837 | 14.55938 | 29.28481 | 9.838849 | 42.10494 | 0.168525 | 1.019652 | 869.9449 |
| 0.3 | 3.022685 | 14.48194 | 29.40561 | 9.839902 | 42.07018 | 0.173259 | 1.006417 | 864.8555 |

- Step 5 is repeated to generate the data given in Tables 7.4–7.9 for pyrolysis fractions of 0.1, 0.3, 0.4, 0.5, 0.7, and 1.0, respectively.

**TABLE 7.4**   Simulation Results With Pyrolysis Fraction of 0.1

| z (m) | xCH₄db | xCO₂db | xCOdb | xH₂db | xN₂db | xO₂db | xTdb | T (K) |
|-------|--------|--------|-------|-------|-------|-------|------|-------|
| 0 | 0.683767 | 2.136513 | 1.884888 | 0.40873 | 73.3777 | 19.4178 | 2.090597 | 1223.213 |
| 0.02 | 0.800868 | 4.796336 | 18.5201 | 2.485898 | 65.25563 | 8.071282 | 0.069883 | 1089.527 |
| 0.04 | 0.782389 | 4.763594 | 23.70237 | 2.432425 | 63.38258 | 4.912288 | 0.024353 | 1061.269 |
| 0.06 | 0.769658 | 4.371855 | 26.76436 | 2.398218 | 62.27495 | 3.406551 | 0.014407 | 1042.676 |
| 0.08 | 0.758898 | 3.963361 | 29.29388 | 2.369227 | 61.36023 | 2.245882 | 0.008517 | 1026.117 |
| 0.1 | 0.74997 | 3.550734 | 31.34968 | 2.344884 | 60.61729 | 1.381913 | 0.005528 | 1011.514 |
| 0.12 | 0.743349 | 3.136718 | 32.83544 | 2.32777 | 60.07947 | 0.872391 | 0.00486 | 999.5503 |
| 0.14 | 0.73844 | 2.774766 | 33.94565 | 2.315215 | 59.6774 | 0.544502 | 0.004027 | 989.3494 |
| 0.16 | 0.736036 | 2.534335 | 34.50031 | 2.309147 | 59.47638 | 0.440589 | 0.003206 | 982.0515 |
| 0.18 | 0.73417 | 2.345318 | 34.92853 | 2.304619 | 59.32099 | 0.363733 | 0.00264 | 975.3781 |
| 0.2 | 0.7327 | 2.194477 | 35.26459 | 2.301203 | 59.19888 | 0.305922 | 0.002231 | 969.1588 |
| 0.22 | 0.731437 | 2.063902 | 35.55259 | 2.298363 | 59.09413 | 0.25768 | 0.001896 | 963.1609 |
| 0.24 | 0.730403 | 1.955809 | 35.78765 | 2.296164 | 59.0085 | 0.219838 | 0.001642 | 957.4216 |
| 0.26 | 0.729461 | 1.856751 | 36.00122 | 2.294218 | 58.93063 | 0.186294 | 0.001421 | 951.7655 |
| 0.28 | 0.728641 | 1.769648 | 36.18686 | 2.292599 | 58.86287 | 0.158148 | 0.001239 | 946.2362 |
| 0.3 | 0.727997 | 1.700617 | 36.33198 | 2.291445 | 58.80977 | 0.137094 | 0.001104 | 940.9216 |

**TABLE 7.5**  Simulation Results With Pyrolysis Fraction of 0.3

| z (m) | xCH$_4$db | xCO$_2$db | xCOdb | xH$_2$db | xN$_2$db | xO$_2$db | xTdb | T (K) |
|---|---|---|---|---|---|---|---|---|
| 0 | 1.793286 | 5.983357 | 4.52639 | 1.193733 | 64.2504 | 16.73459 | 5.518246 | 1216.658 |
| 0.02 | 1.939471 | 10.97722 | 19.32236 | 6.311135 | 54.90821 | 5.584005 | 0.957596 | 1033.664 |
| 0.04 | 1.935614 | 10.82748 | 22.88598 | 6.224867 | 53.84928 | 3.647776 | 0.628997 | 1008.183 |
| 0.06 | 1.926772 | 10.47774 | 25.05023 | 6.166586 | 53.20036 | 2.690798 | 0.487509 | 991.953 |
| 0.08 | 1.918418 | 10.13137 | 26.89201 | 6.115704 | 52.6479 | 1.919906 | 0.374691 | 977.3256 |
| 0.1 | 1.910552 | 9.788829 | 28.43693 | 6.071631 | 52.1843 | 1.319827 | 0.287936 | 964.2793 |
| 0.12 | 1.903667 | 9.453611 | 29.57711 | 6.039214 | 51.84015 | 0.953053 | 0.233199 | 953.6108 |
| 0.14 | 1.897471 | 9.141969 | 30.54866 | 6.011203 | 51.54668 | 0.66448 | 0.189533 | 943.8239 |
| 0.16 | 1.891933 | 8.852876 | 31.3594 | 5.987378 | 51.30158 | 0.450418 | 0.15642 | 934.9296 |
| 0.18 | 1.88729 | 8.595358 | 31.94831 | 5.971027 | 51.12153 | 0.339067 | 0.137417 | 927.3596 |
| 0.2 | 1.883424 | 8.376051 | 32.3834 | 5.958749 | 50.98793 | 0.283237 | 0.127212 | 920.7163 |
| 0.22 | 1.880387 | 8.212385 | 32.70647 | 5.947485 | 50.89048 | 0.241011 | 0.121782 | 914.7475 |
| 0.24 | 1.877616 | 8.04548 | 32.97472 | 5.94267 | 50.8048 | 0.236786 | 0.117933 | 909.0191 |
| 0.26 | 1.875479 | 7.921938 | 33.18175 | 5.937623 | 50.73994 | 0.227883 | 0.115384 | 903.6426 |
| 0.28 | 1.874403 | 7.898341 | 33.33408 | 5.923641 | 50.7029 | 0.154025 | 0.112608 | 898.6482 |
| 0.3 | 1.872802 | 7.839248 | 33.47609 | 5.928906 | 50.64984 | 0.123952 | 0.109166 | 893.4925 |

**TABLE 7.6**  Simulation Results With Pyrolysis Fraction of 0.4

| z (m) | xCH$_4$db | xCO$_2$db | xCOdb | xH$_2$db | xN$_2$db | xO$_2$db | xTdb | T (K) |
|---|---|---|---|---|---|---|---|---|
| 0 | 2.249388 | 7.670138 | 5.497392 | 1.570227 | 60.47409 | 15.60308 | 6.935679 | 1214.557 |
| 0.02 | 2.358232 | 13.32908 | 19.12835 | 7.81675 | 50.9722 | 4.816696 | 1.578694 | 1020.104 |
| 0.04 | 2.362606 | 13.15168 | 22.32043 | 7.725115 | 50.10696 | 3.191278 | 1.141928 | 994.5143 |
| 0.06 | 2.358868 | 12.84356 | 24.22698 | 7.664438 | 49.58503 | 2.383778 | 0.937347 | 978.4768 |
| 0.08 | 2.354802 | 12.54164 | 25.85063 | 7.611405 | 49.14031 | 1.730052 | 0.771162 | 964.0482 |
| 0.1 | 2.35047 | 12.24602 | 27.21302 | 7.565429 | 48.76695 | 1.217988 | 0.640135 | 951.2043 |
| 0.12 | 2.346054 | 11.96113 | 28.21846 | 7.531454 | 48.48943 | 0.899678 | 0.553792 | 940.7342 |
| 0.14 | 2.341909 | 11.69654 | 29.07593 | 7.50211 | 48.25247 | 0.64752 | 0.483522 | 931.1356 |
| 0.16 | 2.338024 | 11.45142 | 29.79191 | 7.477181 | 48.05429 | 0.458561 | 0.428613 | 922.419 |
| 0.18 | 2.334452 | 11.23399 | 30.31369 | 7.459786 | 47.90794 | 0.355565 | 0.394567 | 915.0066 |
| 0.2 | 2.33128 | 11.0477 | 30.70111 | 7.446711 | 47.79857 | 0.300861 | 0.373765 | 908.4938 |
| 0.22 | 2.328706 | 10.9043 | 30.99146 | 7.435189 | 47.71781 | 0.261969 | 0.360557 | 902.6119 |
| 0.24 | 2.326279 | 10.76475 | 31.23439 | 7.429029 | 47.64668 | 0.248486 | 0.350385 | 896.973 |
| 0.26 | 2.324373 | 10.65734 | 31.42475 | 7.423369 | 47.59168 | 0.235693 | 0.342804 | 891.6454 |
| 0.28 | 2.323455 | 10.61589 | 31.5689 | 7.411811 | 47.55741 | 0.186935 | 0.335602 | 886.6034 |
| 0.3 | 2.321369 | 10.48274 | 31.72666 | 7.416828 | 47.50229 | 0.220512 | 0.329607 | 881.3661 |

**TABLE 7.7**   Simulation Results With Pyrolysis Fraction of 0.5

| z (m) | xCH$_4$db | xCO$_2$db | xCOdb | xH$_2$db | xN$_2$db | xO$_2$db | xTdb | T (K) |
|---|---|---|---|---|---|---|---|---|
| 0 | 2.654314 | 9.220438 | 6.3035 | 1.937125 | 57.10507 | 14.5857 | 8.193847 | 1212.836 |
| 0.02 | 2.71126 | 15.30506 | 18.92459 | 9.119592 | 47.55892 | 4.176675 | 2.203893 | 1010.318 |
| 0.04 | 2.721441 | 15.08822 | 21.87514 | 9.023781 | 46.82381 | 2.778434 | 1.689182 | 984.5493 |
| 0.06 | 2.721709 | 14.79636 | 23.60909 | 8.961831 | 46.38768 | 2.086214 | 1.43711 | 968.549 |
| 0.08 | 2.721098 | 14.51514 | 25.08446 | 8.90776 | 46.01644 | 1.524817 | 1.230287 | 954.1726 |
| 0.1 | 2.719706 | 14.2442 | 26.3206 | 8.860972 | 45.70528 | 1.084172 | 1.065063 | 941.3943 |
| 0.12 | 2.717472 | 13.98975 | 27.22973 | 8.826455 | 45.47465 | 0.808254 | 0.953691 | 931.0078 |
| 0.14 | 2.715189 | 13.75467 | 28.00444 | 8.796707 | 45.2778 | 0.589074 | 0.862114 | 921.4931 |
| 0.16 | 2.71286 | 13.5382 | 28.65039 | 8.771516 | 45.11334 | 0.424195 | 0.789489 | 912.8604 |
| 0.18 | 2.710421 | 13.3488 | 29.11936 | 8.753913 | 44.9922 | 0.332627 | 0.742678 | 905.5327 |
| 0.2 | 2.708094 | 13.18722 | 29.46651 | 8.740784 | 44.90179 | 0.283166 | 0.712429 | 899.0959 |
| 0.22 | 2.706164 | 13.06115 | 29.72687 | 8.729518 | 44.8349 | 0.249435 | 0.691968 | 893.2665 |
| 0.24 | 2.704293 | 12.9427 | 29.9442 | 8.722932 | 44.77628 | 0.233905 | 0.675695 | 887.6871 |
| 0.26 | 2.702817 | 12.84983 | 30.11545 | 8.717218 | 44.73052 | 0.221023 | 0.663137 | 882.3965 |
| 0.28 | 2.702187 | 12.80342 | 30.24861 | 8.707302 | 44.70038 | 0.186384 | 0.651718 | 877.33 |
| 0.3 | 2.700592 | 12.70546 | 30.38558 | 8.709663 | 44.65728 | 0.199657 | 0.641771 | 872.1881 |

**TABLE 7.8**   Simulation Results With Pyrolysis Fraction of 0.7

| z (m) | xCH$_4$db | xCO$_2$db | xCOdb | xH$_2$db | xN$_2$db | xO$_2$db | xTdb | T (K) |
|---|---|---|---|---|---|---|---|---|
| 0 | 3.339963 | 11.96311 | 7.546082 | 2.639443 | 51.3492 | 12.83295 | 10.32925 | 1210.284 |
| 0.02 | 3.276033 | 18.41253 | 18.60652 | 11.25691 | 41.92005 | 3.163528 | 3.364431 | 996.9014 |
| 0.04 | 3.292506 | 18.09041 | 21.28155 | 11.15293 | 41.34504 | 2.08564 | 2.751928 | 970.8264 |
| 0.06 | 3.297582 | 17.78331 | 22.81646 | 11.08843 | 41.01267 | 1.562635 | 2.438916 | 954.7727 |
| 0.08 | 3.30121 | 17.49805 | 24.11886 | 11.03243 | 40.73067 | 1.13917 | 2.179615 | 940.3697 |
| 0.1 | 3.303519 | 17.23341 | 25.20583 | 10.98428 | 40.49538 | 0.807694 | 1.96989 | 927.5904 |
| 0.12 | 3.304162 | 17.00019 | 25.9983 | 10.94908 | 40.32245 | 0.60034 | 1.825472 | 917.2441 |
| 0.14 | 3.304381 | 16.7884 | 26.67157 | 10.91893 | 40.17528 | 0.435846 | 1.705593 | 907.7772 |
| 0.16 | 3.304195 | 16.59724 | 27.23046 | 10.89359 | 40.05283 | 0.312439 | 1.609247 | 899.2 |
| 0.18 | 3.303426 | 16.43684 | 27.63096 | 10.87611 | 39.96368 | 0.244035 | 1.544959 | 891.9427 |
| 0.2 | 3.302429 | 16.30333 | 27.92367 | 10.86336 | 39.89789 | 0.207844 | 1.501476 | 885.5758 |
| 0.22 | 3.301569 | 16.19872 | 28.1417 | 10.85282 | 39.84952 | 0.18499 | 1.470687 | 879.7952 |
| 0.24 | 3.300652 | 16.1053 | 28.32182 | 10.84629 | 39.80762 | 0.172731 | 1.44559 | 874.2742 |
| 0.26 | 3.299952 | 16.03128 | 28.46379 | 10.84093 | 39.77478 | 0.163503 | 1.425764 | 869.0213 |
| 0.28 | 3.299857 | 15.98506 | 28.57812 | 10.83283 | 39.75186 | 0.143975 | 1.408293 | 863.9305 |
| 0.3 | 3.299054 | 15.92019 | 28.68686 | 10.83324 | 39.72291 | 0.145224 | 1.392517 | 858.87 |

**TABLE 7.9** Simulation Results With Pyrolysis Fraction of 1.0

| $z$ (m) | $x$CH$_4$db | $x$CO$_2$db | $x$COdb | $x$H$_2$db | $x$N$_2$db | $x$O$_2$db | $x$Tdb | $T$ (K) |
|---|---|---|---|---|---|---|---|---|
| 0 | 4.137064 | 15.34355 | 8.790453 | 3.611764 | 44.54579 | 10.75056 | 12.82081 | 1207.747 |
| 0.02 | 3.889377 | 21.5073 | 18.11428 | 13.65442 | 35.59625 | 2.374664 | 4.863708 | 984.9278 |
| 0.04 | 3.912189 | 21.18074 | 20.51063 | 13.5461 | 35.16957 | 1.515532 | 4.165238 | 958.8356 |
| 0.06 | 3.921534 | 20.87597 | 21.89587 | 13.47882 | 34.92135 | 1.111331 | 3.795133 | 942.5885 |
| 0.08 | 3.928735 | 20.59492 | 23.06362 | 13.42079 | 34.71213 | 0.790369 | 3.489428 | 928.0967 |
| 0.1 | 3.934068 | 20.3375 | 24.03328 | 13.3712 | 34.53856 | 0.543679 | 3.241709 | 915.2811 |
| 0.12 | 3.937075 | 20.11661 | 24.744 | 13.33481 | 34.41028 | 0.389978 | 3.06725 | 904.8459 |
| 0.14 | 3.939324 | 19.91779 | 25.34635 | 13.30375 | 34.30133 | 0.269551 | 2.921904 | 895.3091 |
| 0.16 | 3.94084 | 19.74092 | 25.84229 | 13.2779 | 34.21141 | 0.181559 | 2.805077 | 886.7043 |
| 0.18 | 3.941411 | 19.59607 | 26.19757 | 13.26012 | 34.14588 | 0.134075 | 2.72487 | 879.3992 |
| 0.2 | 3.941454 | 19.47816 | 26.45216 | 13.24756 | 34.09837 | 0.112165 | 2.670127 | 873.024 |
| 0.22 | 3.941439 | 19.38589 | 26.64008 | 13.23736 | 34.06385 | 0.100593 | 2.63079 | 867.2321 |
| 0.24 | 3.941264 | 19.30701 | 26.79202 | 13.2311 | 34.03443 | 0.095396 | 2.598767 | 861.7244 |
| 0.26 | 3.941169 | 19.24456 | 26.91091 | 13.22626 | 34.01144 | 0.092326 | 2.573343 | 856.4771 |
| 0.28 | 3.941562 | 19.20152 | 27.0087 | 13.21888 | 33.99532 | 0.082727 | 2.551296 | 851.3574 |
| 0.3 | 3.941201 | 19.14688 | 27.10178 | 13.21928 | 33.9742 | 0.086188 | 2.530471 | 846.3182 |

The air entering the gasifier contains 79% $N_2$ and 21% $O_2$ by volume. The downdraft gasifiers operated using air contained 45%–50% $N_2$ by volume. $N_2$ can be used to alter the pyrolysis fraction () because it was not produced from any of the reactions. The details of further simulations carried out using the model developed are discussed on our earlier research work (Chaurasia, 2016).

## 7.2 CASE STUDY ON EFFECT OF PARTICLE GEOMETRIES OF BIOMASS IN DOWNDRAFT GASIFIER

This case study proposed a single-particle model of pyrolysis coupled with a more advanced 1D steady-state model of a downdraft gasifier to study the geometrical effects of feedstock on various parameters to establish some of the combustion characteristics of the fuel relevant for gasification. The biomass briquette spheres (0.003–0.05 mm), cylinders (0.003–0.05 mm) with different lengths (10–50 mm) and slab (0.003–0.05 mm) with different lengths (10–50 mm) have been considered. The pyrolysis fractions for different geometries such as slab, cylindrical, and spherical were predicted from the single-particle model. The model was used to analyze parameters such as $d_p$ and $k_B$, moisture content, $T_g$, $W_a$, $X_{O_2}$, and $k_m$ on the performance of the gasification process for different geometries (slab, cylindrical, and spherical). These models were developed in an earlier study by Chaurasia (2018) and are briefly presented in Table 7.10.

**TABLE 7.10**   Single Particle Pyrolysis Model

---

*Kinetic Scheme*

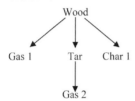

*Particle model*

Mass conservation for biomass, (Gas)$_1$ (Char)$_1$, Tar and (Gas)$_2$:

$$\frac{\partial \bar{C}_B}{\partial \tau} = \left( R^2/\alpha \right)\left( -k_{1(p)}\bar{C}_B - k_{2(p)}\bar{C}_B - k_{3(p)}\bar{C}_B \right) \tag{7.13}$$

$$\frac{\partial \bar{C}_{G1}}{\partial \tau} = \left( R^2/\alpha \right)\left( k_{1(p)}\bar{C}_B \right) \tag{7.14}$$

$$\frac{\partial \bar{C}_{C1}}{\partial \tau} = \left( R^2/\alpha \right)\left( k_{3(p)}\bar{C}_B \right) \tag{7.15}$$

$$\frac{\partial \bar{C}_T}{\partial \tau} = \left( R^2/\alpha \right)\left( k_{2(p)}\bar{C}_B - k_{4(p)}\bar{C}_T \right) \tag{7.16}$$

$$\frac{\partial \bar{C}_{G2}}{\partial \tau} = \left( R^2/\alpha \right)\left( k_{4(p)}\bar{C}_T \right) \tag{7.17}$$

*Enthalpy:*

$$\frac{\partial \theta}{\partial \tau} = \frac{b-1}{x}\frac{\partial \theta}{\partial x} + \frac{\partial^2 \theta}{\partial x^2} + Q\left( -\frac{\partial \rho}{\partial \tau} \right) \tag{7.18}$$

*Initial conditions:*

$$\tau = 0;\ \bar{C}_B = 1, C_{G1} = C_{C1} = C_T = C_{G2} = 0,\ \theta(x,0) = 1 \tag{7.19}$$

*Particle boundary conditions:*

$$\tau > 0;\ \ x = 0,\ \ \ \frac{\partial \theta}{\partial x} = 0 \tag{7.20}$$

---

**TABLE 7.10**  *(Continued)*

$$\tau > 0; \quad x = 1, \quad \frac{\partial \theta}{\partial x} = -\theta B_{iM} \tag{7.21}$$

*Conversion of biomass:*

$$X = \frac{\overline{C}_{B0} - \left[\left(\sum_{i=1}^{M} \overline{C}_B\right) / (M+1)\right]}{\overline{C}_{B0}} \tag{7.22}$$

Dimensionless groups:

$$\alpha = k_B / \rho C_{pB} \tag{7.23}$$

$$x = r' / R \tag{7.24}$$

$$\tau = \alpha t / R^2 \tag{7.25}$$

$$\theta = (T_s - T_f) / (T_0 - T_f) \tag{7.26}$$

$$Bi_M = (R / k_B) \left[ h + \omega \sigma \left( T_s^3 + T_s^2 T_f + T_f^2 T_s + T_f^3 \right) \right] \tag{7.27}$$

$$Q = \left( -\Delta H + C_{pB} T_s \right) / \left( \rho C_{pB} (T_0 - T_f) \right) \tag{7.28}$$

The processes occurring in the downdraft gasifier include: pyrolysis, tar cracking, combustion of gases, and combustion and gasification of char particles, which are described by the model Equations (7.1)–(7.11). Equations (7.13)–(7.21) provide a set of nine coupled partial differential equations for the single-particle model, and Equations (7.1)–(7.11) provide a set of 11 coupled first-order ordinary differential equations for the system variables (where ranges over the eight different species considered), and. These equations along with the initial conditions are solved using COMSOL Multiphysics.

### 7.2.1  SIMULATION APPROACH (SINGLE-PARTICLE PYROLYSIS MODEL)

*Step 1*

- Open COMSOL Multiphysics.
- Select *1D Space Dimension* from the list of options. Hit the *next* arrow at the upper right corner.

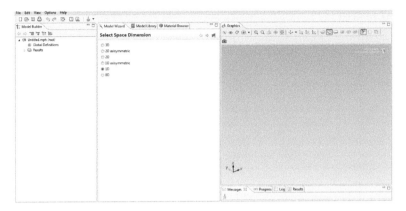

- Select and expand the *Mathematics* folder from the list of options in *Model Wizard*. Further select and expand *PDE Interface* and click on the *Coefficient Form PDE (c)*. Again hit the *next* arrow.

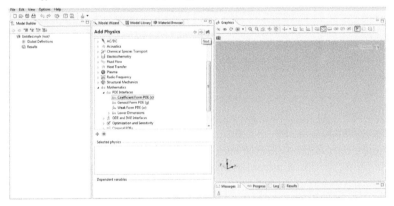

- Then, select *Time Dependent* from the list of *Study Type* options, and click on the *Finish flag* at the upper right corner of the application.

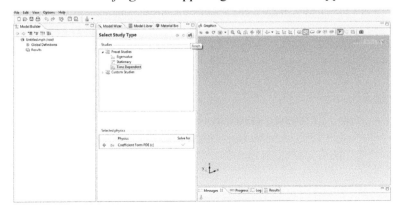

*Step 2*

- In the *Model Builder*, select *Variables* by right clicking on the *Definitions* tab. Another window will open namely *Variables 1*. Click on it and define the variables as given in Table 7.11.

**TABLE 7.11**  Variables and Expressions in the Single Particle Model Equations

| Name | Expression | Description |
|------|-----------|-------------|
| $T_s$ | $T_f - \text{Theta}*(T_f - T_0)$ | Solid temperature ($K$) |
| $k_1$ | $A_1 * \exp(-E_1/(R_c * T_s))$ | Rate constant for reaction 1 (1 s$^{-1}$) |
| $k_2$ | $A_2 * \exp(-E_2/(R_c * T_s))$ | Rate constant for reaction 2 (1 s$^{-1}$) |
| $k_3$ | $A_3 * \exp(-E_3/(R_c * T_s))$ | Rate constant for reaction 3 (1 s$^{-1}$) |
| $k_4$ | $A_4 * \exp(-E_4/(R_c * T_s))$ | Rate constant for reaction 4 (1 s$^{-1}$) |
| $k_B$ | $0.13 + 0.0003*(T_s - 273)$ | Thermal conductivity of biomass (W/m K) |
| $CpB$ | $1112.0 + 4.85*(T_s - 273)$ | Specific heat of biomass (J/kg K) |
| $Q$ | $(-dH + CpB*T_s)/(\text{init\_density}*$ $(CB + CC_1)*CpB*(T_f - T_0))$ | Heat of reaction number (m$^3$/kg) |
| $A_1$ | 1.30E+08 | Arrhenius constant (1 s$^{-1}$) |
| $A_2$ | 2.00E+08 | Arrhenius constant (1 s$^{-1}$) |
| $A_3$ | 1.08E+07 | Arrhenius constant (1 s$^{-1}$) |
| $A_4$ | 6.31E+10 | Arrhenius constant (1 s$^{-1}$) |
| $E_1$ | 140,000 | Activation energy (J/mol) |
| $E_2$ | 133,000 | Activation energy (J/mol) |
| $E_3$ | 121,000 | Activation energy (J/mol) |
| $E_4$ | 185,000 | Activation energy (J/mol) |
| $R_c$ | 8.314 | Ideal gas constant (J/mol K) |
| dim_radius | 1 | Dimensionless radius (–) |
| $dH$ | –255,000 | Heat of reaction (J/kg) |
| $T_0$ | 303 | Room temperature ($K$) |
| Stefan | 5.67E–08 | Stefan–Boltzmann constant (W/m$^2$K$^4$) |
| therm_diff | 1.79E–07 | Thermal diffusivity (m$^2$/s) |
| $b$ | 1 | Geometry factor (–) [slab = 1; cylinder = 2; sphere = 3] |
| init_density | 650 | Initial density of biomass (kg/m$^3$) |
| $T_f$ | 1200 | Final temperature (K) |
| $H$ | 25.2 | Heat transfer coefficient (W/m$^2$ K) |
| Emissivity | 0.95 | emissivity coefficient (–) |
| Radius | 0.009 | Particle radius (m) |
| Time | $(t*\text{radius}^2)/\text{therm\_diff}$ | Time in seconds based on time step (s) |

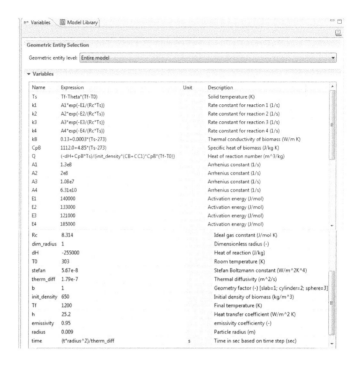

| Name | Expression | Unit | Description |
|------|-----------|------|-------------|
| Ts | Tf-Theta*(Tf-T0) | | Solid temperature (K) |
| k1 | A1*exp(-E1/(Rc*Ts)) | | Rate constant for reaction 1 (1/s) |
| k2 | A2*exp(-E2/(Rc*Ts)) | | Rate constant for reaction 2 (1/s) |
| k3 | A3*exp(-E3/(Rc*Ts)) | | Rate constant for reaction 3 (1/s) |
| k4 | A4*exp(-E4/(Rc*Ts)) | | Rate constant for reaction 4 (1/s) |
| kB | 0.13+0.0003*(Ts-273) | | Thermal conductivity of biomass (W/m K) |
| CpB | 1112.0+4.85*(Ts-273) | | Specific heat of biomass (J/kg K) |
| Q | (-dH+CpB*Ts)/(init_density*(CB+CC1)*CpB*(Tf-T0)) | | Heat of reaction number (m^3/kg) |
| A1 | 1.3e8 | | Arrhenius constant (1/s) |
| A2 | 2e8 | | Arrhenius constant (1/s) |
| A3 | 1.08e7 | | Arrhenius constant (1/s) |
| A4 | 6.31e10 | | Arrhenius constant (1/s) |
| E1 | 140000 | | Activation energy (J/mol) |
| E2 | 133000 | | Activation energy (J/mol) |
| E3 | 121000 | | Activation energy (J/mol) |
| E4 | 185000 | | Activation energy (J/mol) |
| Rc | 8.314 | | Ideal gas constant (J/mol K) |
| dim_radius | 1 | | Dimensionless radius (-) |
| dH | -255000 | | Heat of reaction (J/kg) |
| T0 | 303 | | Room temperature (K) |
| stefan | 5.67e-8 | | Stefan Boltzmann constant (W/m^2K^4) |
| therm_diff | 1.79e-7 | | Thermal diffusivity (m^2/s) |
| b | 1 | | Geometry factor (-) [slab=1; cylinder=2; sphere=3] |
| init_density | 650 | | Initial density of biomass (kg/m^3) |
| Tf | 1200 | | Final temperature (K) |
| h | 25.2 | | Heat transfer coefficient (W/m^2 K) |
| emissivity | 0.95 | | emissivity coefficiency (-) |
| radius | 0.009 | | Particle radius (m) |
| time | (t*radius^2)/therm_diff | s | Time in sec based on time step (sec) |

## Step 3

- With the *Model Builder*, right click on the *Geometry* option. From the list of options, click on the *Interval* taskbar.
- In the *Interval* environment, select *Intervals*1, *Left endpoint: 0 and Right endpoint: 1*. Click on the *Build All* option at the upper part of tool bar. A straight line graph will appear on the right side of the application window.

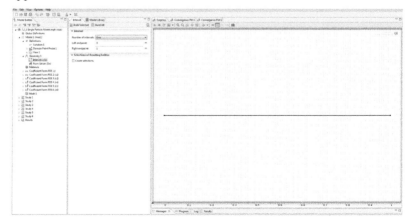

*Step 4 (Define the Problem for $C_B$)*

- In the *Model Builder*, select *Coefficient Form PDE* node and change the Dependent variables to CB.

- Now, in the *Model Builder*, select *Coefficient Form PDE 1* option. This will open a tab to enter the coefficients of the characteristics of a model equation.
- In the *Domain Selection* panel, you will see an equation of the form

$$e_a \frac{\partial^2 CB}{\partial t^2} + d_a \frac{\partial CB}{\partial t} + \nabla \cdot \left( -c\nabla CB - \alpha CB + \gamma \right) + \beta \cdot \nabla CB + aCB = f \quad (7.29)$$

where $\nabla = \left[ \dfrac{\partial}{\partial x} \right]$.

In order to solve the governing differential equation, we need to assign the coefficients in above equation a suitable value.

To convert Equation (7.29) to the desired form of Equation (7.13), the value of coefficients in Equation (7.29) to be changed as follows:

$c = 0$ $\alpha = 0$, $\beta = 0$, $\gamma = 0$
$a = 0$ $f = (((radius^2)/therm\_diff) * (-k1 * CB - k_2 * CB - k_3 * CB))$
$e_a = 0$ $d_a = 1$.

This adjustment will reduce Equation (7.29) to Equation (7.13) format.

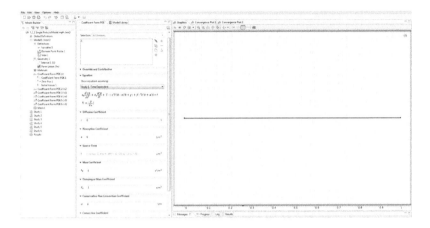

- Now, go to *Initial Values 1* available in the *Coefficient Form PDE* tab and set *CB* to 1. This will add the initial condition given in Equation (7.19).

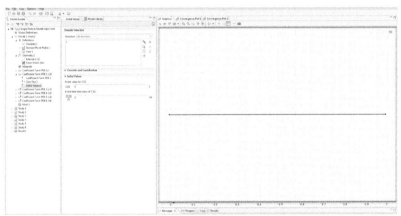

*Step 5 (Define the Problem for $C_{G1}$)*

- Next, to add the equation for $C_{G1}$, right click on Model 1 and select the *Mathematics* folder from the list of options in *Model Wizard*. Further select and expand *PDE Interface* and click on the *Coefficient Form PDE (c)*. Again hit the *next* arrow.
- Then, select *Time Dependent* from the list of *Study Type* options, and click on the *Finish flag* at the upper right corner of the application.
- In the *Model Builder*, select *Coefficient Form PDE 2* node and change the Dependent variables to CG1.

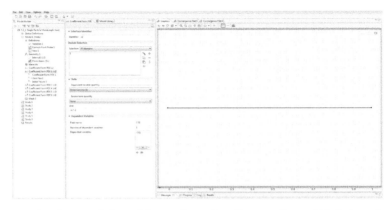

- Now, in the *Model Builder*, select *Coefficient Form PDE 1* option under *Coefficient Form PDE 2*. This will open a tab to enter the coefficients of the characteristics of a model equation.
- In the *Domain Selection* panel, you will see an equation of the form

$$e_a \frac{\partial^2 CG1}{\partial t^2} + d_a \frac{\partial CG1}{\partial t} + \nabla \cdot \left(-c\nabla CG1 - \alpha CG1 + \gamma\right) + \beta \cdot \nabla CG1 + aCG1 = f \quad (7.30)$$

where $\nabla = \left[\dfrac{\partial}{\partial x}\right]$.

In order to solve the governing differential equation, we need to assign the coefficients in above equation a suitable value.

To convert Equation (7.30) to the desired form of Equation (7.14), the value of coefficients in Equation (7.30) to be changed as follows:

$c = 0 \; \alpha = 0, \; \beta = 0, \; \gamma = 0$

$a = 0 \; f = (((radius^2)/therm\_diff)*(k_1*CB))$

$e_a = 0 \; d_a = 1.$

This adjustment will reduce Equation (7.30) to Equation (7.14) format.

- Now, go to *Initial Values 1* available in the *Coefficient Form PDE 2* tab and set *CG1* to 0. This will add the initial condition given in Equation (7.19).

## Step 6 (Define the Problem for $C_{C1}$, $C_1$, $C_{G2}$)

- Repeat Step 5 to define the problems for $C_{C1}$, $C_1$, $C_{G2}$ with initial values of $C_{C1}$, $C_1$, $C_{G2}$ set to zero.

## Step 7 (Define the Problem for θ)

- Next, to add the equation for θ, right click on Model 1 and select the *Mathematics* folder from the list of options in *Model Wizard*. Further select and expand *PDE Interface* and click on the *Coefficient Form PDE (c)*. Again hit the *next* arrow.
- Then select *Time Dependent* from the list of *Study Type* options, and click on the *Finish flag* at the upper right corner of the application.
- In the *Model Builder*, select *Coefficient Form PDE 6* node and change the Dependent variable to Theta.

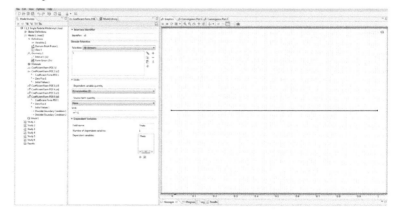

- Now, in the *Model Builder*, select *Coefficient Form PDE 1* option under *Coefficient Form PDE 6*. This will open a tab to enter the coefficients of the characteristics of a model equation.
- In the *Domain Selection* panel, you will see an equation of the form

$$e_a \frac{\partial^2 Theta}{\partial t^2} + d_a \frac{\partial Theta}{\partial t} + \nabla \cdot \left( -c\nabla Theta - \alpha Theta + \gamma \right)$$
$$+ \beta \cdot \nabla Theta + aTheta = f \tag{7.31}$$

where $\nabla = \left[ \dfrac{\partial}{\partial x} \right]$.

In order to solve the governing differential equation, we need to assign the coefficients in above equation a suitable value.

To convert Equation (7.31) to the desired form of Equation (7.18), the value of coefficients in Equation (7.31) to be changed as follows:

$c = 1$ $\alpha = 0$, $\beta = (1-b)/x$, $\gamma = 0$

$a = 0$ $f = (Q*(radius^2)*(k_1*CB+k_2*CB))/(therm\_diff)$

$e_a = 0$ $d_a = 1$.

This adjustment will reduce Equation (7.31) to Equation (7.18) format.

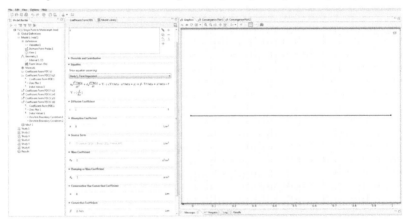

### Step 8

- Right click on the *Coefficient Form PDE 6* tab in *Model Builder*. Select the *Dirichlet Boundary Condition* option. Another window will open namely *Dirichlet Boundary Condition 1*. Click on it.
- Select the left point on the horizontal line graph. Click on the "+" sign at the top right corner, this will add *boundary 1* in the *Boundary Selection* tab. At this point, put *r: Theta+Thetax*. This step will add a *Neumann boundary condition*: $at\ x = 0, \dfrac{\partial \theta}{\partial x} = 0$, that is, $\dfrac{\partial Theta}{\partial x} = 0$, that is, *Thetax = 0*

as given in Equation (7.20) (Note: as *Theta = r, r = Theta + Thetax, Thetax = 0*).

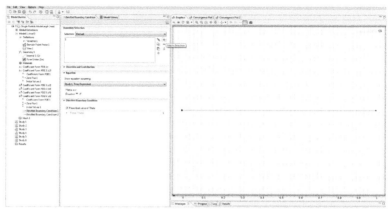

- Again right click on the *Coefficient Form PDE 6 tab* in *Model Builder*. Select the *Dirichlet Boundary Condition* option. Another window will open namely *Dirichlet Boundary Condition 2*. Click on it.
- Select the right point on the horizontal line graph. Click on the "+" sign at the top right corner, this will add *boundary 2* in the *Boundary Selection* tab. At this point, put *r: Theta+Thetax+Theta\*(radius/k$_B$)\*(h+emissiv ity\*stefan\*(T$_s^3$+(T$_s^2$)\*T$_f$+(T$_f^2$)\*T$_s$+T$_f^3$))*. This step will add a *Neumann boundary condition:* $at\ x = 1, \dfrac{\partial\theta}{\partial x} = -\theta Bi_M$, that is, $\dfrac{\partial Theta}{\partial x} = -Theta * Bi_M$, that is, thetax $= -$ Theta\* $Bi_M$ as given in Equation (7.21) (Note: as *Theta = r, r = Theta + Thetax + Theta\*(radius/k$_B$)\*(h + emissivity\*stefan\* (T$_s^3$+(T$_s^2$)\*T$_f$+(T$_f^2$)\*T$_s$+T$_f^3$)); Thetax+Theta\*(radius/k$_B$)\*(h+emissivity\* stefan\*(T$_s^3$+(T$_s^2$)\*T$_f$+(T$_f^2$)\*T$_s$+T$_f^3$)) = 0; Bi$_M$* is given by equation (7.27).

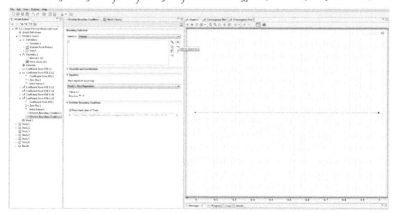

- Now, go to *Initial Values 1* available in the *Coefficient Form PDE 6* tab and set *Theta* to 1. This will add the initial condition given in Equation (7.19).

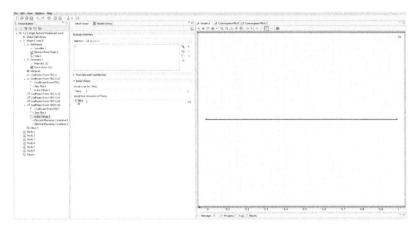

*Step 9*

- Click on the *Mesh* option in *Model Builder*. Select *Normal Mesh* Type. Click on *Build All* option at the top of ribbon. A dialogue box will appear in the *Message* tab as: *"Complete mesh consists of 15 elements."*

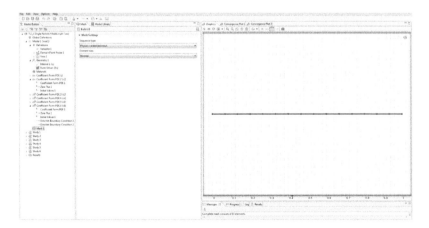

- Now, go to *Study* option in the model pellet tab. Click on the *Compute* (=) button. Save the simulation.

*Step 10*

- The single particle model is simulated to analyze the fractions of gas, tar and char from pyrolysis for several thermal conductivities of biomass with slab, cylinder and sphere. Consider radius of particle as 0.009 m and time step as 0.5 sec then dimensionless time = (therm_ diff*time step)/radius$^2$ = 0.001104938. To continue simulation for 300 s, we have to write range as: range (0,0.001104938,0.662962963) [If 0.001104938*600 = 0.662962963].

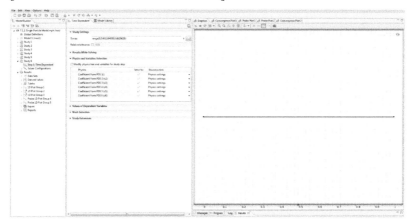

- To simulate the model for thermal conductivity of slab as 0.05 W/m K, take $b = 1$ and $k_B = 0.05$ W/m K and capture the results of simulation using "Domain Point Probe." To add it, right click on "Definitions"; then select "Probes" and then "Domain Point Probe." Then, right click on "Domain Point Probe" and select "Point Probe Expression" for different variables. The probe setting for variable $CB$ is shown in Figure 7.1.

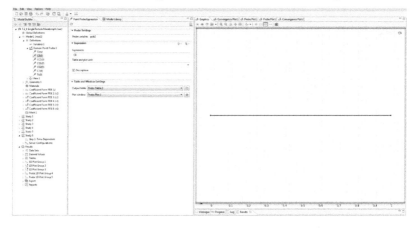

- The result of the simulation is given in Table 7.12. The result at dimensionless time of 0.434014 is considered as the concentration of biomass (9.90E−05) is negligible at this point. Similarly, simulations have been carried out for cylinder and sphere geometry considering $b = 2$ and $b = 3$, respectively, for $k_B = 0.05$ W/m K. The same procedure is repeated for $k_B = 0.1, 0.2, 0.3, 0.35, 0.4$ W/m K for different geometries. The results are summarized in Table 7.13.

**TABLE 7.12** Simulation Results for Slab Geometry ($b = 1$, $k_B = 0.05$ W/m K)

| Dimensionless time | Time (s) | $CB$ | $CC_1$ | $CG_1$ | $CG_2$ | $CT$ | $T_s$ |
|---|---|---|---|---|---|---|---|
| 0 | 0 | 1 | 9.61E−16 | 2.35E−17 | 1.55E−23 | 2.53E−16 | 303 |
| 5.59E−05 | 0.025311 | 1 | 1.43E−15 | 5.59E−17 | 2.44E−22 | 4.87E−16 | 303 |
| 1.12E−04 | 0.050622 | 1 | 2.03E−15 | 1.40E−16 | 2.01E−21 | 9.97E−16 | 303 |
| 2.24E−04 | 0.101243 | 1 | 3.85E−15 | 6.09E−16 | 6.86E−20 | 3.55E−15 | 303 |
| 3.52E−04 | 0.159277 | 1 | 8.59E−15 | 2.65E−15 | 7.05E−18 | 1.36E−14 | 303 |
| 4.80E−04 | 0.217311 | 1 | 1.80E−14 | 7.66E−15 | 2.41E−17 | 3.71E−14 | 303 |
| 6.08E−04 | 0.275346 | 1 | 3.53E−14 | 1.84E−14 | 1.45E−16 | 8.56E−14 | 303 |
| 7.24E−04 | 0.327576 | 1 | 5.94E−14 | 3.47E−14 | 5.49E−16 | 1.57E−13 | 303 |
| 8.39E−04 | 0.379807 | 1 | 9.39E−14 | 5.98E−14 | 1.86E−15 | 2.63E−13 | 303 |
| 9.55E−04 | 0.432038 | 1 | 1.39E−13 | 9.45E−14 | 5.08E−15 | 4.06E−13 | 303 |
| 0.00107 | 0.484269 | 1 | 1.94E−13 | 1.39E−13 | 1.26E−14 | 5.84E−13 | 303 |
| 0.001186 | 0.5365 | 1 | 2.58E−13 | 1.93E−13 | 2.72E−14 | 7.88E−13 | 303 |
| 0.001301 | 0.58873 | 1 | 3.26E−13 | 2.52E−13 | 5.45E−14 | 1.00E−12 | 303 |
| 0.001416 | 0.640961 | 1 | 3.95E−13 | 3.15E−13 | 9.97E−14 | 1.20E−12 | 303 |
| 0.001647 | 0.745423 | 1 | 5.16E−13 | 4.29E−13 | 2.69E−13 | 1.48E−12 | 303 |
| 0.001878 | 0.849884 | 1 | 5.92E−13 | 5.08E−13 | 5.67E−13 | 1.48E−12 | 303 |
| 0.002109 | 0.954346 | 1 | 6.19E−13 | 5.45E−13 | 9.72E−13 | 1.21E−12 | 303 |
| 0.00234 | 1.058807 | 1 | 6.09E−13 | 5.51E−13 | 1.41E−12 | 7.71E−13 | 303 |
| 0.002571 | 1.163269 | 1 | 5.73E−13 | 5.33E−13 | 1.79E−12 | 2.99E−13 | 303 |
| 0.002804 | 1.269072 | 1 | 5.17E−13 | 5.01E−13 | 2.05E−12 | −1.16E−13 | 303 |
| 0.003015 | 1.364295 | 1 | 4.45E−13 | 4.55E−13 | 2.19E−12 | −4.66E−13 | 303 |
| 0.003225 | 1.459518 | 1 | 3.50E−13 | 3.92E−13 | 2.26E−12 | −8.22E−13 | 303 |
| 0.003395 | 1.53649 | 1 | 2.57E−13 | 3.27E−13 | 2.28E−12 | −1.13E−12 | 303 |
| 0.003566 | 1.613462 | 1 | 1.49E−13 | 2.49E−13 | 2.28E−12 | −1.47E−12 | 303 |
| 0.003736 | 1.690433 | 1 | 2.65E−14 | 1.58E−13 | 2.28E−12 | −1.86E−12 | 303 |
| 0.003906 | 1.767405 | 1 | −1.08E−13 | 5.59E−14 | 2.26E−12 | −2.28E−12 | 303 |
| 0.004076 | 1.844377 | 1 | −2.51E−13 | −5.66E−14 | 2.25E−12 | −2.74E−12 | 303 |

**TABLE 7.12**   *(Continued)*

| Dimensionless time | Time (s) | CB | $CC_1$ | $CG_1$ | $CG_2$ | CT | $T_s$ |
|---|---|---|---|---|---|---|---|
| 0.004246 | 1.921349 | 1 | −4.04E−13 | −1.81E−13 | 2.20E−12 | −3.21E−12 | 303 |
| 0.004416 | 1.998321 | 1 | −5.56E−13 | −3.08E−13 | 2.14E−12 | −3.68E−12 | 303 |
| 0.004756 | 2.152264 | 1 | −8.31E−13 | −5.52E−13 | 1.94E−12 | −4.46E−12 | 303 |
| 0.005048 | 2.284157 | 1 | −1.03E−12 | −7.38E−13 | 1.69E−12 | −4.95E−12 | 303 |
| 0.005339 | 2.416049 | 1 | −1.17E−12 | −8.87E−13 | 1.34E−12 | −5.18E−12 | 303 |
| 0.005601 | 2.534753 | 1 | −1.24E−12 | −9.78E−13 | 9.40E−13 | −5.11E−12 | 303 |
| 0.005838 | 2.641586 | 1 | −1.27E−12 | −1.03E−12 | 5.22E−13 | −4.88E−12 | 303 |
| 0.006074 | 2.748419 | 1 | −1.26E−12 | −1.05E−12 | 3.78E−14 | −4.43E−12 | 303 |
| 0.00631 | 2.855251 | 1 | −1.21E−12 | −1.04E−12 | −4.95E−13 | −3.80E−12 | 303 |
| 0.006546 | 2.962084 | 1 | −1.12E−12 | −1.00E−12 | −1.06E−12 | −3.05E−12 | 303 |
| 0.006782 | 3.068917 | 1 | −1.01E−12 | −9.43E−13 | −1.63E−12 | −2.18E−12 | 303 |
| 0.007254 | 3.282583 | 1 | −7.02E−13 | −7.47E−13 | −2.78E−12 | −1.40E−13 | 303 |
| 0.007726 | 3.496249 | 1 | −2.90E−13 | −4.67E−13 | −3.82E−12 | 2.14E−12 | 303 |
| 0.008198 | 3.709915 | 1 | 2.08E−13 | −1.10E−13 | −4.65E−12 | 4.53E−12 | 303 |
| 0.008671 | 3.923581 | 1 | 7.71E−13 | 3.10E−13 | −5.22E−12 | 6.89E−12 | 303 |
| 0.009143 | 4.137247 | 1 | 1.37E−12 | 7.75E−13 | −5.49E−12 | 9.12E−12 | 303 |
| 0.009615 | 4.350912 | 1 | 1.97E−12 | 1.26E−12 | −5.52E−12 | 1.12E−11 | 303 |
| 0.010087 | 4.564578 | 1 | 2.53E−12 | 1.73E−12 | −5.31E−12 | 1.29E−11 | 303 |
| 0.010559 | 4.778244 | 1 | 3.00E−12 | 2.16E−12 | −4.89E−12 | 1.42E−11 | 303 |
| 0.011032 | 4.99191 | 1 | 3.36E−12 | 2.49E−12 | −4.26E−12 | 1.49E−11 | 303 |
| 0.011504 | 5.205576 | 1 | 3.56E−12 | 2.72E−12 | −3.42E−12 | 1.49E−11 | 303 |
| 0.012448 | 5.632907 | 1 | 3.49E−12 | 2.78E−12 | −1.08E−12 | 1.26E−11 | 303 |
| 0.01328 | 6.009314 | 1 | 2.89E−12 | 2.38E−12 | 1.66E−12 | 8.11E−12 | 303 |
| 0.014112 | 6.385721 | 1 | 1.92E−12 | 1.65E−12 | 4.78E−12 | 1.87E−12 | 303 |
| 0.014943 | 6.762128 | 1 | 6.62E−13 | 6.46E−13 | 8.04E−12 | −5.57E−12 | 303 |
| 0.015775 | 7.138535 | 1 | −6.74E−13 | −4.51E−13 | 1.10E−11 | −1.31E−11 | 303 |
| 0.016607 | 7.514942 | 1 | −2.02E−12 | −1.60E−12 | 1.34E−11 | −2.01E−11 | 303.0001 |
| 0.017356 | 7.853708 | 1 | −3.11E−12 | −2.58E−12 | 1.47E−11 | −2.54E−11 | 303.0001 |
| 0.018104 | 8.192475 | 1 | −3.98E−12 | −3.41E−12 | 1.53E−11 | −2.92E−11 | 303.0002 |
| 0.018853 | 8.531241 | 1 | −4.56E−12 | −4.03E−12 | 1.49E−11 | −3.13E−11 | 303.0004 |
| 0.019602 | 8.870007 | 1 | −4.79E−12 | −4.37E−12 | 1.38E−11 | −3.13E−11 | 303.0007 |
| 0.02035 | 9.208773 | 1 | −4.64E−12 | −4.41E−12 | 1.18E−11 | −2.94E−11 | 303.0011 |
| 0.021099 | 9.54754 | 1 | −4.11E−12 | −4.14E−12 | 9.26E−12 | −2.54E−11 | 303.0017 |
| 0.021848 | 9.886306 | 1 | −3.20E−12 | −3.55E−12 | 6.11E−12 | −1.97E−11 | 303.0025 |
| 0.022596 | 10.22507 | 1 | −1.96E−12 | −2.67E−12 | 2.60E−12 | −1.24E−11 | 303.0036 |
| 0.023345 | 10.56384 | 1 | −4.60E−13 | −1.55E−12 | −1.23E−12 | −3.76E−12 | 303.0052 |

**TABLE 7.12** *(Continued)*

| Dimensionless time | Time (s) | CB | $CC_1$ | $CG_1$ | $CG_2$ | CT | $T_s$ |
|---|---|---|---|---|---|---|---|
| 0.024093 | 10.9026 | 1 | 1.22E−12 | −2.45E−13 | −5.22E−12 | 5.67E−12 | 303.0074 |
| 0.024842 | 11.24137 | 1 | 2.99E−12 | 1.17E−12 | −9.23E−12 | 1.56E−11 | 303.0102 |
| 0.025591 | 11.58014 | 1 | 4.79E−12 | 2.65E−12 | −1.31E−11 | 2.55E−11 | 303.0138 |
| 0.027088 | 12.25767 | 1 | 7.98E−12 | 5.41E−12 | −1.97E−11 | 4.32E−11 | 303.0247 |
| 0.028435 | 12.86745 | 1 | 1.01E−11 | 7.46E−12 | −2.37E−11 | 5.53E−11 | 303.0394 |
| 0.029783 | 13.47723 | 1 | 1.12E−11 | 8.73E−12 | −2.53E−11 | 6.16E−11 | 303.0604 |
| 0.031131 | 14.08701 | 1 | 1.11E−11 | 9.10E−12 | −2.44E−11 | 6.15E−11 | 303.0893 |
| 0.032478 | 14.69679 | 1 | 9.75E−12 | 8.48E−12 | −2.08E−11 | 5.47E−11 | 303.1276 |
| 0.033826 | 15.30657 | 1 | 7.30E−12 | 6.93E−12 | −1.49E−11 | 4.16E−11 | 303.1768 |
| 0.035173 | 15.91634 | 1 | 4.09E−12 | 4.66E−12 | −7.07E−12 | 2.38E−11 | 303.2391 |
| 0.036521 | 16.52612 | 1 | 4.98E−13 | 1.92E−12 | 2.06E−12 | 2.98E−12 | 303.3165 |
| 0.037868 | 17.1359 | 1 | −2.96E−12 | −9.28E−13 | 1.18E−11 | −1.86E−11 | 303.4113 |
| 0.039216 | 17.74568 | 1 | −5.77E−12 | −3.49E−12 | 2.15E−11 | −3.86E−11 | 303.5257 |
| 0.041911 | 18.96524 | 1 | −7.99E−12 | −6.48E−12 | 3.79E−11 | −6.60E−11 | 303.8214 |
| 0.044336 | 20.06284 | 1 | −4.81E−12 | −5.54E−12 | 4.61E−11 | −6.81E−11 | 304.1748 |
| 0.046762 | 21.16045 | 1 | 2.92E−12 | −1.12E−12 | 4.62E−11 | −4.74E−11 | 304.6218 |
| 0.049188 | 22.25805 | 1 | 1.40E−11 | 6.28E−12 | 3.75E−11 | −5.69E−12 | 305.1723 |
| 0.051613 | 23.35565 | 1 | 2.59E−11 | 1.49E−11 | 2.16E−11 | 4.73E−11 | 305.8344 |
| 0.053796 | 24.34349 | 1 | 3.45E−11 | 2.16E−11 | 4.12E−12 | 9.29E−11 | 306.5324 |
| 0.055979 | 25.33134 | 1 | 3.96E−11 | 2.61E−11 | −1.34E−11 | 1.29E−10 | 307.33 |
| 0.058162 | 26.31918 | 1 | 4.05E−11 | 2.76E−11 | −2.84E−11 | 1.48E−10 | 308.23 |
| 0.060345 | 27.30702 | 1 | 3.74E−11 | 2.59E−11 | −3.84E−11 | 1.50E−10 | 309.2345 |
| 0.062528 | 28.29486 | 1 | 3.13E−11 | 2.14E−11 | −4.18E−11 | 1.33E−10 | 310.3449 |
| 0.064711 | 29.28271 | 1 | 2.40E−11 | 1.52E−11 | −3.80E−11 | 1.04E−10 | 311.5616 |
| 0.066894 | 30.27055 | 1 | 1.76E−11 | 8.78E−12 | −2.68E−11 | 6.65E−11 | 312.8848 |
| 0.069077 | 31.25839 | 1 | 1.47E−11 | 3.81E−12 | −8.96E−12 | 3.08E−11 | 314.3138 |
| 0.07126 | 32.24623 | 1 | 1.73E−11 | 1.89E−12 | 1.39E−11 | 4.62E−12 | 315.8475 |
| 0.073443 | 33.23408 | 1 | 2.68E−11 | 4.26E−12 | 3.91E−11 | −4.24E−12 | 317.4841 |
| 0.075626 | 34.22192 | 1 | 4.35E−11 | 1.15E−11 | 6.40E−11 | 8.74E−12 | 319.2212 |
| 0.077809 | 35.20976 | 1 | 6.67E−11 | 2.33E−11 | 8.64E−11 | 4.42E−11 | 321.057 |
| 0.079992 | 36.1976 | 1 | 9.46E−11 | 3.87E−11 | 1.04E−10 | 9.94E−11 | 322.9889 |
| 0.082175 | 37.18544 | 1 | 1.25E−10 | 5.62E−11 | 1.16E−10 | 1.69E−10 | 325.0132 |
| 0.084358 | 38.17329 | 1 | 1.55E−10 | 7.41E−11 | 1.20E−10 | 2.48E−10 | 327.1274 |
| 0.086541 | 39.16113 | 1 | 1.84E−10 | 9.08E−11 | 1.16E−10 | 3.27E−10 | 329.3284 |
| 0.088724 | 40.14897 | 1 | 2.09E−10 | 1.05E−10 | 1.06E−10 | 4.01E−10 | 331.613 |
| 0.09309 | 42.12466 | 1 | 2.46E−10 | 1.21E−10 | 7.33E−11 | 5.12E−10 | 336.4161 |

**TABLE 7.12**   *(Continued)*

| Dimensionless time | Time (s) | CB | $CC_1$ | $CG_1$ | $CG_2$ | CT | $T_s$ |
|---|---|---|---|---|---|---|---|
| 0.097456 | 44.10034 | 1 | 2.74E−10 | 1.23E−10 | 3.44E−11 | 5.77E−10 | 341.5131 |
| 0.101822 | 46.07603 | 1 | 3.17E−10 | 1.22E−10 | 6.23E−12 | 6.31E−10 | 346.8791 |
| 0.106188 | 48.05171 | 1 | 4.05E−10 | 1.34E−10 | 1.48E−12 | 7.37E−10 | 352.489 |
| 0.110554 | 50.0274 | 1 | 5.61E−10 | 1.70E−10 | 2.64E−11 | 9.37E−10 | 358.3223 |
| 0.11492 | 52.00308 | 1 | 8.17E−10 | 2.40E−10 | 8.24E−11 | 1.28E−09 | 364.354 |
| 0.119286 | 53.97876 | 1 | 1.21E−09 | 3.48E−10 | 1.64E−10 | 1.80E−09 | 370.5597 |
| 0.123652 | 55.95445 | 1 | 1.79E−09 | 4.84E−10 | 2.58E−10 | 2.47E−09 | 376.9174 |
| 0.128018 | 57.93013 | 1 | 2.67E−09 | 6.38E−10 | 3.50E−10 | 3.32E−09 | 383.4075 |
| 0.132384 | 59.90582 | 1 | 4.12E−09 | 8.05E−10 | 4.24E−10 | 4.46E−09 | 390.0088 |
| 0.13675 | 61.8815 | 1 | 6.51E−09 | 1.00E−09 | 4.73E−10 | 6.09E−09 | 396.7092 |
| 0.141117 | 63.85719 | 1 | 1.07E−08 | 1.29E−09 | 4.93E−10 | 8.77E−09 | 403.4936 |
| 0.145483 | 65.83287 | 1 | 1.80E−08 | 1.77E−09 | 4.88E−10 | 1.35E−08 | 410.3472 |
| 0.149849 | 67.80856 | 1 | 3.09E−08 | 2.62E−09 | 4.71E−10 | 2.20E−08 | 417.2567 |
| 0.154215 | 69.78424 | 1 | 5.34E−08 | 4.14E−09 | 4.59E−10 | 3.72E−08 | 424.2102 |
| 0.158581 | 71.75993 | 1 | 9.23E−08 | 6.85E−09 | 4.76E−10 | 6.41E−08 | 431.1971 |
| 0.162947 | 73.73561 | 1 | 1.59E−07 | 1.16E−08 | 5.44E−10 | 1.11E−07 | 438.2077 |
| 0.167313 | 75.7113 | 1 | 2.70E−07 | 2.01E−08 | 6.85E−10 | 1.94E−07 | 445.2333 |
| 0.171679 | 77.68698 | 0.999999 | 4.55E−07 | 3.48E−08 | 9.10E−10 | 3.36E−07 | 452.2658 |
| 0.176045 | 79.66267 | 0.999999 | 7.68E−07 | 6.13E−08 | 1.22E−09 | 5.88E−07 | 459.2967 |
| 0.180411 | 81.63835 | 0.999998 | 1.28E−06 | 1.07E−07 | 1.60E−09 | 1.01E−06 | 466.3207 |
| 0.184777 | 83.61404 | 0.999996 | 2.08E−06 | 1.85E−07 | 2.02E−09 | 1.72E−06 | 473.3323 |
| 0.189143 | 85.58972 | 0.999993 | 3.35E−06 | 3.16E−07 | 2.46E−09 | 2.88E−06 | 480.3262 |
| 0.193509 | 87.56541 | 0.999989 | 5.31E−06 | 5.31E−07 | 2.88E−09 | 4.75E−06 | 487.2979 |
| 0.197875 | 89.54109 | 0.999983 | 8.30E−06 | 8.82E−07 | 3.26E−09 | 7.71E−06 | 494.2428 |
| 0.202241 | 91.51677 | 0.999973 | 1.28E−05 | 1.44E−06 | 3.61E−09 | 1.24E−05 | 501.1573 |
| 0.206607 | 93.49246 | 0.999959 | 1.94E−05 | 2.31E−06 | 3.97E−09 | 1.94E−05 | 508.0383 |
| 0.215339 | 97.44383 | 0.999905 | 4.30E−05 | 5.75E−06 | 4.97E−09 | 4.63E−05 | 521.6869 |
| 0.224071 | 101.3952 | 0.999789 | 9.16E−05 | 1.36E−05 | 6.96E−09 | 1.06E−04 | 535.1661 |
| 0.232803 | 105.3466 | 0.999561 | 1.84E−04 | 3.01E−05 | 1.07E−08 | 2.25E−04 | 548.4572 |
| 0.241535 | 109.2979 | 0.999124 | 3.54E−04 | 6.33E−05 | 1.64E−08 | 4.58E−04 | 561.5474 |
| 0.250267 | 113.2493 | 0.998257 | 6.77E−04 | 1.34E−04 | 2.59E−08 | 9.32E−04 | 574.4312 |
| 0.258126 | 116.8055 | 0.99682 | 0.001191 | 2.56E−04 | 4.41E−08 | 0.001732 | 585.8424 |
| 0.265985 | 120.3618 | 0.994444 | 0.002015 | 4.68E−04 | 8.85E−08 | 0.003074 | 597.075 |
| 0.273843 | 123.918 | 0.990616 | 0.003301 | 8.21E−04 | 2.13E−07 | 0.005262 | 608.1252 |
| 0.281702 | 127.4742 | 0.984599 | 0.005261 | 0.001397 | 5.75E−07 | 0.008742 | 618.9905 |
| 0.289561 | 131.0305 | 0.975485 | 0.00815 | 0.002299 | 1.60E−06 | 0.014066 | 629.6689 |

**TABLE 7.12** *(Continued)*

| Dimensionless time | Time (s) | *CB* | *CC₁* | *CG₁* | *CG₂* | *CT* | *Tₛ* |
|---|---|---|---|---|---|---|---|
| 0.29742 | 134.5867 | 0.96192 | 0.01233 | 0.003683 | 4.42E−06 | 0.022063 | 640.1586 |
| 0.305279 | 138.1429 | 0.942167 | 0.018254 | 0.005759 | 1.19E−05 | 0.033808 | 650.458 |
| 0.313138 | 141.6992 | 0.914145 | 0.02644 | 0.008788 | 3.08E−05 | 0.050597 | 660.566 |
| 0.320996 | 145.2554 | 0.875532 | 0.037437 | 0.013073 | 7.68E−05 | 0.073881 | 670.4817 |
| 0.325472 | 147.2807 | 0.84799 | 0.045142 | 0.016186 | 1.24E−04 | 0.090557 | 676.0424 |
| 0.329948 | 149.306 | 0.815445 | 0.054095 | 0.019928 | 2.09E−04 | 0.110323 | 681.5407 |
| 0.333976 | 151.1288 | 0.78205 | 0.063177 | 0.023811 | 3.18E−04 | 0.130644 | 686.4358 |
| 0.338004 | 152.9516 | 0.744615 | 0.073229 | 0.028218 | 4.89E−04 | 0.153448 | 691.2805 |
| 0.342032 | 154.7743 | 0.703158 | 0.084245 | 0.03315 | 7.29E−04 | 0.178718 | 696.0751 |
| 0.34606 | 156.5971 | 0.657788 | 0.096155 | 0.03861 | 0.001084 | 0.206363 | 700.8202 |
| 0.350088 | 158.4199 | 0.60709 | 0.109271 | 0.044797 | 0.001633 | 0.237209 | 705.492 |
| 0.353424 | 159.9295 | 0.563298 | 0.120505 | 0.050184 | 0.002226 | 0.263787 | 709.3303 |
| 0.356761 | 161.4391 | 0.517741 | 0.13208 | 0.055839 | 0.003011 | 0.29133 | 713.1344 |
| 0.360097 | 162.9487 | 0.471335 | 0.143777 | 0.061642 | 0.004008 | 0.319238 | 716.9095 |
| 0.363433 | 164.4583 | 0.424396 | 0.1555 | 0.067562 | 0.005298 | 0.347244 | 720.6529 |
| 0.366769 | 165.9679 | 0.377595 | 0.167123 | 0.073495 | 0.006817 | 0.37497 | 724.377 |
| 0.370105 | 167.4776 | 0.331461 | 0.178496 | 0.079384 | 0.008688 | 0.401972 | 728.0764 |
| 0.376777 | 170.4968 | 0.244581 | 0.199665 | 0.090591 | 0.013968 | 0.451194 | 735.4012 |
| 0.382782 | 173.2141 | 0.175099 | 0.216348 | 0.099674 | 0.020846 | 0.488033 | 741.9175 |
| 0.388186 | 175.6596 | 0.122228 | 0.22888 | 0.106665 | 0.029238 | 0.512989 | 747.7288 |
| 0.392922 | 177.8025 | 0.084435 | 0.237731 | 0.111715 | 0.038635 | 0.527483 | 752.7834 |
| 0.397183 | 179.7311 | 0.057592 | 0.243956 | 0.115334 | 0.049005 | 0.534114 | 757.3043 |
| 0.400966 | 181.4429 | 0.03923 | 0.248175 | 0.117828 | 0.059899 | 0.534868 | 761.2939 |
| 0.404749 | 183.1548 | 0.02541 | 0.251328 | 0.119718 | 0.072556 | 0.530989 | 765.2591 |
| 0.408101 | 184.6712 | 0.016614 | 0.253317 | 0.120929 | 0.0853 | 0.52384 | 768.7499 |
| 0.411452 | 186.1877 | 0.010351 | 0.254724 | 0.121797 | 0.099576 | 0.513553 | 772.2175 |
| 0.414254 | 187.4557 | 0.006755 | 0.255526 | 0.122298 | 0.112721 | 0.5027 | 775.0981 |
| 0.417056 | 188.7237 | 0.004241 | 0.256084 | 0.12265 | 0.126997 | 0.490028 | 777.9601 |
| 0.419578 | 189.8649 | 0.0027 | 0.256423 | 0.122867 | 0.140821 | 0.477189 | 780.5195 |
| 0.4221 | 191.0061 | 0.001637 | 0.256657 | 0.123017 | 0.155567 | 0.463122 | 783.063 |
| 0.42437 | 192.0332 | 0.00102 | 0.256792 | 0.123104 | 0.169621 | 0.449463 | 785.3384 |
| 0.426639 | 193.0603 | 6.22E−04 | 0.256878 | 0.123161 | 0.184401 | 0.434938 | 787.6009 |
| 0.428682 | 193.9847 | 3.85E−04 | 0.256929 | 0.123195 | 0.198302 | 0.421189 | 789.6259 |
| 0.430521 | 194.8166 | 2.46E−04 | 0.256959 | 0.123215 | 0.211284 | 0.408296 | 791.4394 |
| 0.432359 | 195.6485 | 1.54E−04 | 0.256979 | 0.123228 | 0.224687 | 0.394952 | 793.2444 |
| 0.434014 | 196.3973 | 9.90E−05 | 0.256991 | 0.123236 | 0.237091 | 0.382583 | 794.8617 |

**TABLE 7.13** Fractions of Gas, Tar, and Char From Pyrolysis for Several Thermal Conductivities of Biomass With Slab, Cylinder, and Sphere ($dp = 0.018$ m, $T_0 = 303$ K, $T_f = 1228$ K)

| $k_B$ | GasSlab | TarSlab | CharSlab | GasCyl | TarCyl | CharCyl | GasSph | TarSph | CharSph |
|---|---|---|---|---|---|---|---|---|---|
| 0.05 | 0.360327 | 0.382583 | 0.256991 | 0.447223 | 0.308952 | 0.243757 | 0.490644 | 0.273428 | 0.235856 |
| 0.1 | 0.380214 | 0.362008 | 0.25771 | 0.435975 | 0.319717 | 0.244243 | 0.497006 | 0.266587 | 0.236348 |
| 0.2 | 0.373027 | 0.368197 | 0.258714 | 0.433146 | 0.32153 | 0.245232 | 0.474511 | 0.288022 | 0.237394 |
| 0.3 | 0.363241 | 0.377117 | 0.259555 | 0.412516 | 0.340979 | 0.246406 | 0.470255 | 0.291205 | 0.238454 |
| 0.35 | 0.359509 | 0.380154 | 0.260269 | 0.423633 | 0.329393 | 0.246887 | 0.4795 | 0.281205 | 0.239225 |
| 0.4 | 0.352821 | 0.386306 | 0.2608 | 0.428364 | 0.324107 | 0.247449 | 0.472751 | 0.28748 | 0.239704 |

## 7.2.2   SIMULATION APPROACH (DOWNDRAFT GASIFIER MODEL)

*Step 1*

- Select *0D Space Dimension* from the list of options by opening the COMSOL Multiphysics. Hit the *next* arrow at the upper right corner.

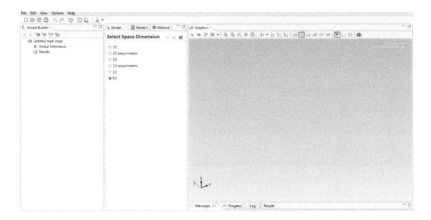

- Select and expand the *Mathematics* folder from the list of options in *Model Wizard*. Further choose *Global ODEs and DAEs* and hit *next tab*.

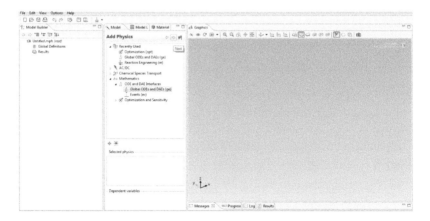

- Then, select *Time Dependent* from the list of *Study Type* options, and click on the *Finish flag* at the upper right corner of the application.

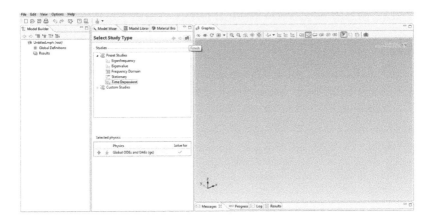

*Step 2*

- Now in the *Model Builder*, select *Global Equations* option. This will open a tab to enter the coefficients of the characteristics of a model equation.
- In the *Global Equations* panel, you will see an equation of the form

$$f\left(u,u_t,u_{tt},t\right) = 0, u\left(t_0\right) = u_0, u_t\left(t_0\right) = u_{t0}. \tag{7.32}$$

To convert Equations (7.1)–(7.11) to the desired form of Equation (7.32), enter the values as given in Table 7.1.

*Step 3*

- In the *Model Builder*, select *Variables* by right clicking on the *Definitions* tab. Another window will open namely *Variables 1*. Click on it and define the variables as given in Table 7.2.

*Step 4*

- Now, expand the *Study 1* tab in *Model Builder*. Select the *Step 1: Time Dependent*. Another window will open namely *Time Dependent*. Expand the *Study Settings* tab and set the range *Times: range (0,0.05,0.6)*.

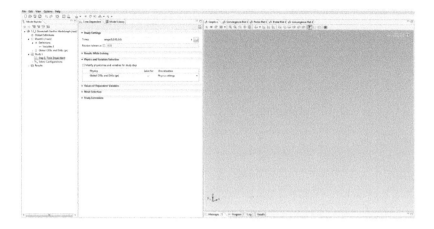

- Click on the *Compute* (=) button. Save the simulation.

## Step 5

- Now, expand the *Study 1* tab in *Model Builder*. Select the *Step 1: Time Dependent*. Another window will open namely *Time Dependent*. Expand the *Study Settings* tab and set the range *Times: range (0,0.05,0.6)*.
- To study the effect of thermal conductivity for different geometries simulations has been carried out using a downdraft gasifier model. Using the experimental data on gas and tar fractions and simulation results of single particle model for $k_B = 0.05$ W/m K, simulations has been performed using downdraft gasifier model given in Table 7.1. The results are reported in Tables 7.14 and 7.15.

**TABLE 7.14** Simulation Results Using a Downdraft Gasifier Model for $k_B = 0.05$ W/m K

| Length (m) | Parameters | Slab | Cylinder | Sphere |
|---|---|---|---|---|
| 0.6 | $xCH_4db$ | 2.07430641 | 2.047512 | 2.053597 |
| 0.6 | $xCO_2db$ | 16.06469825 | 11.64102 | 9.410922 |
| 0.6 | $xCOdb$ | 18.78763344 | 28.29711 | 31.96042 |
| 0.6 | $xH_2db$ | 5.995197099 | 6.802116 | 7.174638 |
| 0.6 | $xN_2db$ | 55.21695965 | 51.11155 | 49.34402 |
| 0.6 | $xO_2db$ | 1.508211654 | 0.031652 | 0.03849 |
| 0.6 | $xTdb$ | 0.352993495 | 0.069046 | 0.017919 |
| 0.6 | $T\,(K)$ | 887.5205506 | 889.5182 | 890.5913 |
| 0.6 | hsg | 3.818868764 | 4.119472 | 4.26755 |
| 0.6 | $P\,(Pa)$ | 102,503.2166 | 102,520.4 | 102,529.3 |

**TABLE 7.15** Simulation Results Using a Downdraft Gasifier Model for a Wide Range of Thermal Conductivities for a Length of 0.6 m

| $k_B$ | $CH_4$slab | $CO_2$slab | COslab | $H_2$slab | $CH_4$cyl | $CO_2$cyl | COcyl | $H_2$cyl | $CH_4$sph | $CO_2$sph | COsph | $H_2$sph |
|---|---|---|---|---|---|---|---|---|---|---|---|---|
| 0.1 | 2.054337 | 14.98192 | 21.38116 | 6.187645 | 2.053508 | 12.37892 | 27.0374 | 6.709572 | 2.050776 | 9.02726 | 32.58777 | 7.222831 |
| 0.2 | 2.057142 | 15.32507 | 20.56607 | 6.116677 | 2.052243 | 12.50434 | 26.80195 | 6.683386 | 2.057221 | 10.39241 | 30.31217 | 7.052232 |
| 0.3 | 2.062739 | 15.79196 | 19.40409 | 6.018849 | 2.06537 | 13.69061 | 24.32208 | 6.508827 | 2.056403 | 10.61153 | 29.93277 | 7.01714 |
| 0.4 | 2.067648 | 16.23756 | 18.19146 | 5.910997 | 2.049413 | 12.5851 | 26.3692 | 6.639923 | 2.051236 | 10.35675 | 30.33535 | 7.031747 |
| 0.05 | 2.074306 | 16.0647 | 18.78763 | 5.995197 | 2.047512 | 11.64102 | 28.29711 | 6.802116 | 2.053597 | 9.410922 | 31.96042 | 7.174638 |
| 0.35 | 2.063602 | 15.94309 | 18.99494 | 5.980028 | 2.054855 | 12.93666 | 25.73783 | 6.602165 | 2.049304 | 9.944368 | 31.03743 | 7.083615 |

The details of further simulations carried using the model developed is discussed in recent research work (Chaurasia, 2018).

## 7.3 ESTIMATION OF KINETIC PARAMETERS FOR SAW DUST

### 7.3.1 PROBLEM STATEMENT

This model shows how to use the Parameter Estimation feature in the Reaction Engineering physics interface to find the Arrhenius parameters of a first-order reaction.

**Note:** This model requires the Optimization Module.

### 7.3.2 MODEL DESCRIPTION

In this case study, the kinetic scheme used for the modeling of primary pyrolysis of sawdust, rice husk and sugarcane bagasse consists of the two competing parallel reactions giving gaseous volatiles and solid char plus unreacted biomass as follows:

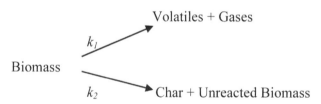

The rate of formation of individual product is modeled as first order in the difference between ultimate yield of the product and the amount of that product generated up to that time [9, 13]. Hence, the kinetic equation of irreversible, first-order pyrolysis reaction is expressed as

$$\frac{dV_i}{dt} = k_i \left( V_i^* - V_i \right). \tag{7.33}$$

Initially, at $t = 0$, $V_i = 0$ for gaseous volatiles and as $t \to \infty$, $V_i \to V_i^*$. The unknown parameter $k_i$ has been the focus of kinetic studies. The temperature-dependent reaction rate of the pyrolysis is often described by the Arrhenius equation as follows:

$$k_i = A_i \, e^{\left( \frac{-E_i}{RT} \right)}. \tag{7.34}$$

Substitution of Equation (7.34) in Equation (7.33) gives

$$\frac{dV_i}{dt} = A_i \ e^{\left(\frac{-E_i}{RT}\right)}\left(V_i^* - V_i\right). \tag{7.35}$$

In Equation (7.35), $A$ is the frequency factor (1/s) and $E$ is the activation energy (J/mol). In order to evaluate the Arrhenius parameters, $A$ and $E$, a set of experiments were conducted (Khonde and Chaurasia, 2015) to determine the volatiles plus gases as function of time for the temperatures; $T = 773$ K, 823 K, 873 K, and 923 K.

The frequency factor $A$ is defined by Equation (7.37), so that the model experiences similar sensitivity with respect to changes in parameter values

$$k = \exp\left(A_{\mathrm{ex}}\right)\exp\left(-\frac{E}{R_g T}\right) \tag{7.36}$$

The frequency factor $A$ is then evaluated as

$$A = \exp\left(A_{\mathrm{ex}}\right). \tag{7.37}$$

The initial guess for the activation energy is 130 kJ/mol and for $A_{ex}$ are 11.5.


### 7.3.3  SIMULATION APPROACH

#### Step 1

- Open COMSOL Multiphysics.
- Select *0D Space Dimension* from the list of options. Hit the *next* arrow at the upper right corner.

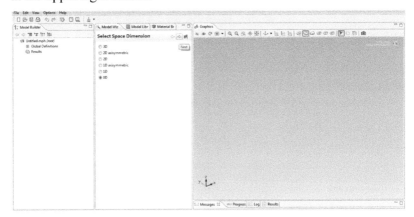

- Expand the *Chemical Species Transport* folder from the list of options in *Model Wizard* and select *Reaction Engineering*. Hit the *next* arrow again.

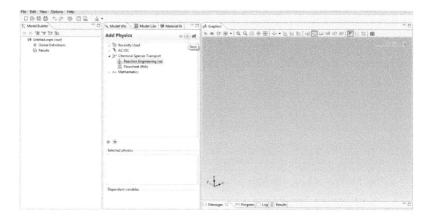

- Then, select *Time Dependent* from the list of *Study Type* options, and click on the *Finish flag*.

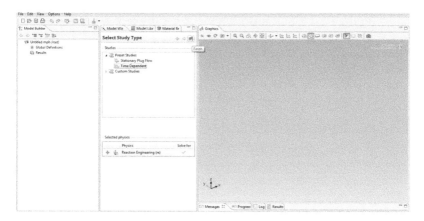

*Step 2*

- With the *Model Builder*, right click on the *Global Definitions* tab and select *Parameters*. Another window will open namely *Parameters*. Click on it and define $T_{iso}$: 773 K.

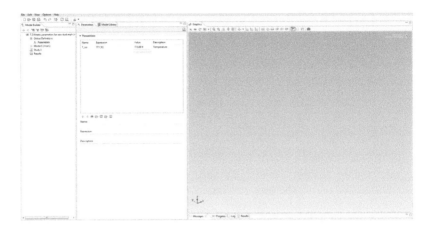

- With the *Model Builder*, right click on the *Reaction Engineering* tab and select the *Reaction* option. Another window will open namely *Reaction 1*. Click on it.
- Select *Reaction 1* option available in the *Reaction Engineering* tab and in the formula edit window type, type $B \Rightarrow VG$. (Note: => Equal to followed by greater sign).
- In the Rate Constants section, check the box "Use Arrhenius expressions" and set $A^f$: $\exp(A_{ex})$, $n^f$: 0 and $E^f$: E.

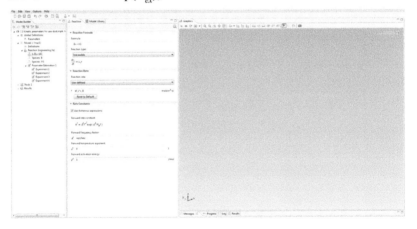

- Select *Species: B* available in the *Reaction Engineering* tab and in the *General Expression* section type, $c_0$: *100* and set *Rate expression: User defined*.

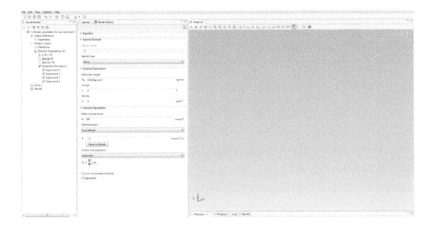

- Select *Species: VG* available in the *Reaction Engineering* tab and in the *General Expression* section type, $c_0$:$0$ and set *Rate expression* to *User defined R: kf_1\*(75.2 − c_VG)*.

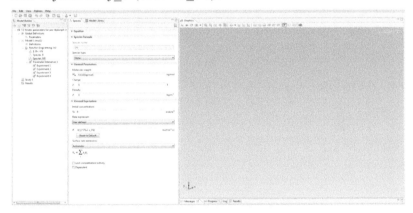

*Step 3*

- With the *Model Builder*, again right click on the *Reaction Engineering* tab in *Model Builder* and select the *Parameter Estimation* option. Another window will open namely *Parameter Estimation 1*. Click on it and enter the following.

| Variable | Initial Value | Lower Bound | Upper Bound |
|----------|---------------|-------------|-------------|
| $A_{ex}$ | 9.2 | | |
| $E$ | $100e_3$ | | |

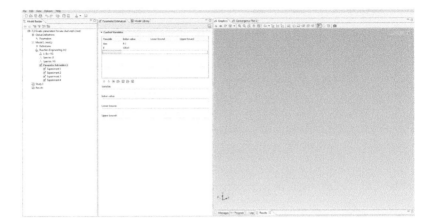

- Right click on the *Parameter Estimation 1* tab in *Model Builder* and select *Experiment* option. Another window will open namely *Experiment 1*.
- Generate the file activation_energy_experiment773K.csv as *comma separated value* files (*csv*-files) with the data given in Table 7.16. Browse the file and click on *Import* button.

**TABLE 7.16**   Experimental Data at 773 K

| Time | conc VG (773 K) |
|------|-----------------|
| 60   | 1.2             |
| 120  | 5               |
| 180  | 27              |
| 240  | 42.8            |
| 300  | 52              |
| 360  | 66.4            |
| 420  | 68.2            |
| 480  | 71.4            |

- In the table, set the Model variables *time: t* and *conc VG* (773 K): *c_VG*.
- In the Experimental Parameters section, click on "+" button at the bottom and enter *Parameter names: $T_{iso}$* and *Parameter expressions: 773*.

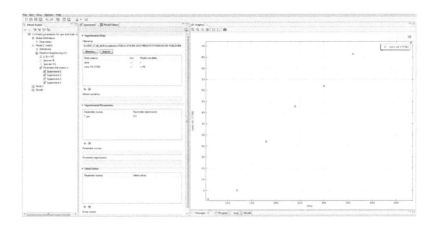

*Step 4*

- Again right click on the *Parameter Estimation 1* tab in *Model Builder* and select *Experiment* option. Another window will open namely *Experiment 2*.
- Generate the file activation_energy_experiment823K.csv as *comma separated value* files (*csv*-files) with the data given in Table 7.17. Browse the file and click on *Import* button.

**TABLE 7.17** Experimental Data at 823 K

| Time | conc VG (823 K) |
|------|-----------------|
| 60   | 1.4  |
| 120  | 11   |
| 180  | 25.2 |
| 240  | 39.8 |
| 300  | 56.6 |
| 360  | 69.8 |
| 420  | 75   |
| 480  | 75   |

- In the table, set the Model variables *time: t* and *conc VG (823 K): c_VG*.
- In the Experimental Parameters section, click on "+" button at the bottom and enter *Parameter names: $T_{iso}$* and *Parameter expressions: 823*.

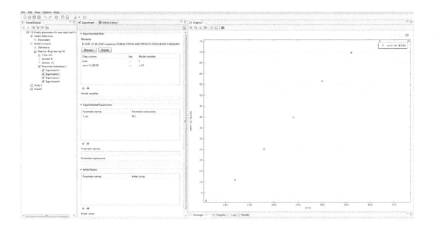

## Step 5

- Again right click on the *Parameter Estimation 1* tab in *Model Builder* and select *Experiment* option. Another window will open namely *Experiment 3*.
- Generate the file activation_energy_experiment873K.csv as *comma separated value* files (*csv*-files) with the data given in Table 7.18. Browse the file and click on *Import* button.

**TABLE 7.18**   Experimental Data at 873 K

| Time | conc VG (873 K) |
|------|-----------------|
| 60   | 2.6             |
| 120  | 9.4             |
| 180  | 42.6            |
| 240  | 67              |
| 300  | 71.8            |
| 360  | 73.4            |
| 420  | 75              |
| 480  | 75              |

- In the table, set the Model variables *time: t* and *conc VG (873 K): c_VG*.
- In the Experimental Parameters section, click on "+" button at the bottom and enter *Parameter names: $T_{iso}$* and *Parameter expressions: 873*.

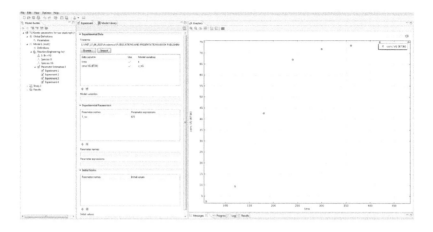

## Step 6

- Again right click on the *Parameter Estimation 1* tab in *Model Builder* and select *Experiment* option. Another window will open namely *Experiment 4*.
- Generate the file activation_energy_experiment923K.csv as *comma separated value* files (*csv*-files) with the data given in Table 7.19. Browse the file and click on *Import* button.

**TABLE 7.19**  Experimental Data at 923 K

| Time | conc VG (923 K) |
|------|-----------------|
| 60   | 3.4             |
| 120  | 27.6            |
| 180  | 58              |
| 240  | 73.2            |
| 300  | 74.6            |
| 360  | 75.2            |
| 420  | 75.2            |
| 480  | 75.2            |

- In the table, set the Model variables *time: t* and *conc VG* (923 K)*: c_VG*.
- In the Experimental Parameters section, click on "+" button at the bottom and enter *Parameter names: $T_{iso}$* and *Parameter expressions: 923*.

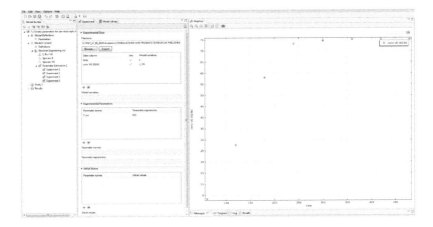

## Step 7

- In the *Reaction Engineering* tab in *Model Builder*, expand the *General section* and set $T: T_{iso}$.

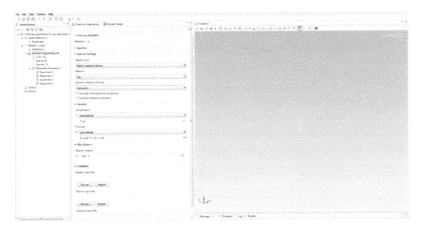

## Step 8

- Now, go to *Study 1, Step1: Time Dependent* option in the model pellet tab and set the *Times: range (0,50,500)*.

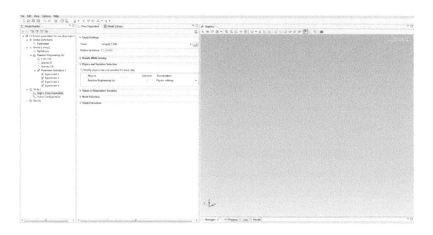

*Step 9*

- Right click on the *Study 1* tab in *Model Builder* and select *Show Default Solver* option. Go to *Solver 1* under *Solver Configurations*. Right click on it and select *Optimization Solver 1*.
- Then right click on *Optimization Solver 1* node, and select *Time-Dependent Solver 1*. Remove *Time-Dependent Solver 1* if it is present under Solver 1.
- In the *Optimization Solver 1* window, from the *Method* list, choose *Levenberg–Marquardt*.

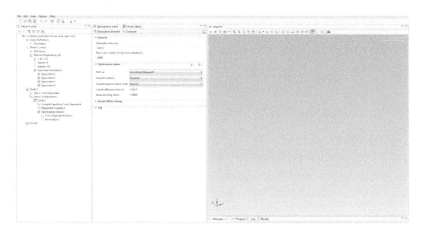

*Step 10*

- Go to *Time-Dependent Solver 1* window under *Optimization Solver 1*. Expand the *Absolute tolerance* section and choose *Global method:*

*Unscaled* and *Tolerance: 1e–5*. Expand the *Output* section and choose *Times to store: Specified values*. Click on the *Compute* (=) button.

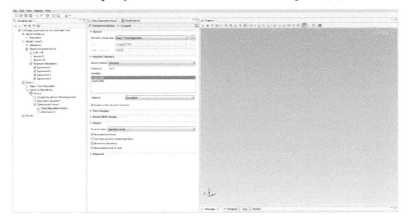

## Step 11

- In the *Model Builder*, expand *Experiment 1 Group node* available under *Results* and select *Global 1* option. In the *Global 1* window, expand the Data section. Set *Parameter selection ($T_{iso}$): From list* and *Parameter values ($T_{iso}$): 773*.
- In the *y*-axis data section, select the *Expression: mod1.re.c_VG* and in the *x*-axis data section, select the select the *Axis source data: Time*. Click on the *Plot* at the top.

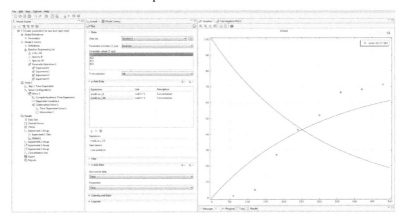

## Step 12

- In the *Model Builder*, expand *Experiment 1 Group node* available under *Results* and select *Global 1* option. In the *Global 1* window, expand the Data section. Set *Parameter selection (T_{iso}): From list* and *Parameter values (T_{iso}): 823.*

- In the *y*-axis data section, select the *Expression: mod1.re.c_VG* and in the *x*-axis data section, select the select the *Axis source data: Time*. Click on the *Plot* at the top.

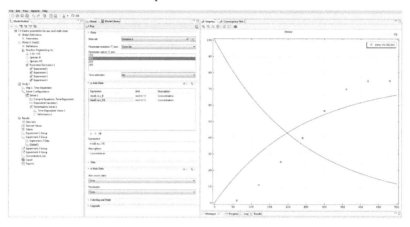

## Step 13

- In the *Model Builder*, expand *Experiment 1 Group node* available under *Results* and select *Global 1* option. In the *Global 1* window, expand the Data section. Set *Parameter selection (T_{iso}): From list* and *Parameter values (T_{iso}): 873.*
- In the *y*-axis data section, select the *Expression: mod1.re.c_VG* and in the *x*-axis data section, select the select the *Axis source data: Time*. Click on the *Plot* at the top.

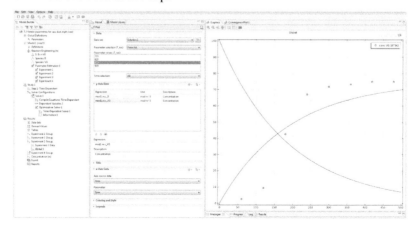

*Step 14*

- In the *Model Builder*, expand *Experiment 1 Group node* available under *Results* and select *Global 1* option. In the *Global 1* window, expand the Data section. Set *Parameter selection ($T_{iso}$): From list* and *Parameter values ($T_{iso}$): 923*.
- In the *y*-axis data section, select the *Expression: mod1.re.c_VG* and in the *x*-axis data section, select the select the *Axis source data: Time*. Click on the *Plot* at the top.

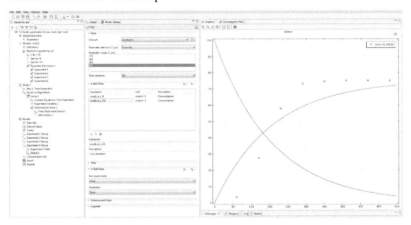

*Step 15*

- In the *Results tab*, right click on *Derived Values* and click *Evaluate All*. $E$ is found to be 25715.05317 and $A_{ex}$ is evaluated to $-1.69468$.
  $A = \exp(A_{ex})$
  $A = \exp(-1.69468) = 0.18366$.

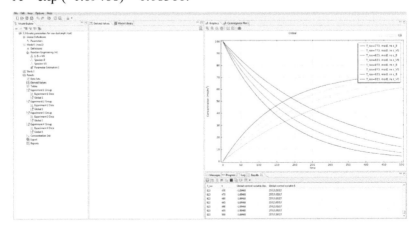

# References

Ahuja, P. *Introduction to Numerical Methods in Chemical Engineering*. PHI Learning: New Delhi, 2010.

Bird, R. B.; Armstrong, R. C.; Hassager, O. *Dynamics of Polymeric Liquids*. Wiley-Interscience: New York, 1987.

Bird, R. B.; Stewart, W. E.; Lightfoot, E. N. *Transport Phenomena*, 2nd ed. Wiley: New York, 2002.

Carnahan, B.; Luther, H. A.; Wilkes, J. O. *Applied Numerical Methods*. Wiley: New York, 1969.

Chapra, S.C.; Canale, R.P. *Numerical Methods for Engineers*, 3rd ed. Tata McGraw-Hill: New Delhi, 1998.

Chaurasia, A. Modeling of downdraft gasifier: Studies on chemical kinetics and operating conditions on the performance of the biomass gasification process. *Energy*, 2016, **116**, 1065–1076.

Chaurasia, A. Modeling of downdraft gasification process: Studies on particle geometries in thermally thick regime. *Energy*, 2018, 142, 991–1009.

Edgar, T. F.; Himmelblau, D. M.; Ladson, L.S. *Optimization of Chemical Processes*. McGraw Hill: New York, 2001.

Finlayson, B. A. *Nonlinear Analysis in Chemical Engineering*. McGraw-Hill: New York, 1980.

Finlayson, B. A. *Numerical Methods for Problems with Moving Fronts*. Ravenna Park: Seattle, WA, 1992.

Finlayson, B. A. *Introduction to Chemical Engineering Computing*. Wiley: New York, 2006.

Fogler, H. S. *Elements of Chemical Reaction Engineering*, 4th ed. Pearson Education/Prentice-Hall: Hoboken, NJ, 2006.

Ghoshdastidar, P. S. *Computer Simulation of Flow and Heat Transfer*. Tata McGraw Hill: New Delhi, 1998.

Gupta, S. K. *Numerical Methods for Engineers*. Wiley Eastern: New Delhi, 1995.

Khonde, R.D.; Chaurasia, A. S. Pyrolysis of sawdust, rice husk and sugarcane bagasse: Kinetic modelling and estimation of kinetic parameters using different optimization tools. *J. Inst. Eng. (India): Series E*, 2015, 96, 23–30.

Levenspiel, O. *Chemical Reaction Engineering*, 3rd ed. Wiley: New York, 2001.

Rhee, H. K.; Aris, R.; Amundson, N. R. *First-Order Partial Differential Equations, Vol. I, Theory and Application of Single Equations*. Prentice-Hall: Hoboken, NJ, 1986.

Schlichting, H. *Boundary-Layer Theory*. McGraw-Hill: New York, 1979.

# Index